Mihai Duduta
Soft Robotics

Also of Interest

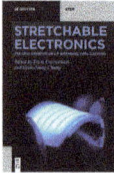

Stretchable Electronics
The Next Generation of Emerging Applications
Edited by Tricia Carmichael, Hyun-Joong Chung, 2025
ISBN 978-3-11-075718-7, e-ISBN (PDF) 978-3-11-075728-6

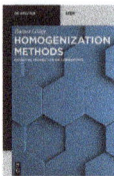

Homogenization Methods
Effective Properties of Composites
Rainer Glüge, 2023
ISBN 978-3-11-079351-2, e-ISBN (PDF) 978-3-11-079352-9

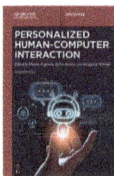

Personalized Human-Computer Interaction
Mirjam Augstein, Eelco Herder, Wolfgang Wörndl (Eds.), 2023
ISBN 978-3-11-099960-0, e-ISBN (PDF) 978-3-11-098856-7

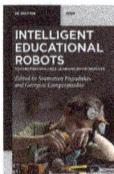

Intelligent Educational Robots
Toward Personalized Learning Environments
Edited by Stamatios Papadakis, Georgios Lampropoulos, 2025
ISBN 978-3-11-135206-0, e-ISBN (PDF) 978-3-11-135269-5

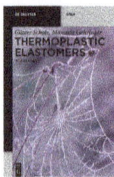

Thermoplastic Elastomers
At a Glance
Günter Scholz, Manuela Gehringer, 2021
ISBN 978-3-11-073983-1, e-ISBN (PDF) 978-3-11-073984-8

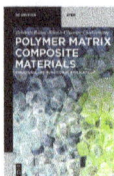

Polymer Matrix Composite Materials
Structural and Functional Applications
Debdatta Ratna, Bikash Chandra Chakraborty, 2024
ISBN 978-3-11-078148-9, e-ISBN (PDF) 978-3-11-078157-1

Mihai Duduta

Soft Robotics

Building Machines from Soft Matter

DE GRUYTER

Author
Prof. Mihai Duduta
University of Connecticut
School of Mechanical, Aerospace, and
Manufacturing Engineering
191 Auditorium Road
Storrs
Connecticut, 06269
United States of America

ISBN 978-3-11-106940-1
e-ISBN (PDF) 978-3-11-106941-8
e-ISBN (EPUB) 978-3-11-106970-8

Library of Congress Control Number: 2025935578

Bibliographic information published by the Deutsche Nationalbibliothek
The Deutsche Nationalbibliothek lists this publication in the Deutsche Nationalbibliografie;
detailed bibliographic data are available on the Internet at http://dnb.dnb.de.

© 2025 Walter de Gruyter GmbH, Berlin/Boston, Genthiner Straße 13, 10785 Berlin
Cover image: photo taken by Mihai Duduta
Typesetting: VTeX UAB, Lithuania

www.degruyter.com
Questions about General Product Safety Regulation:
productsafety@degruyterbrill.com

To Patricia, who reminds me to just keep swimming.

Preface

This book is a product of necessity: as a teacher for a course on Soft Robotics, I have not been able to find a book that enables a novice in the field to quickly grasp the basics. When setting out to write the book, I sought a framework that helps bridge understanding between the highly diverse ideas that fall under the growing interdisciplinary field of Soft Robotics. For those less familiar with the terms, Soft Robots represent a class of machines that are made of materials as soft as living systems. The result of using these building blocks is unique capabilities that allow soft robots to operate in broad range of environments inaccessible to rigid robots, including unstructured environments (e. g., extreme environments, disaster zones), in the proximity of people (e. g., collaborative manufacturing spaces, home companions, and healthcare robotics), and beyond. With Soft Matter being a vast, interdisciplinary, and growing field, the machines made from these building blocks are becoming more complex in how they are made and how they are operated.

The common theme for this book is **energy**: how is energy harvested, stored, transformed, and delivered by soft machines to enable robotic capabilities? By focusing on energy as a common language across different technologies, the book enables the reader to understand the fundamentals of each approach, and to compare robotic capabilities between them. As a result, this is not a book focused heavily on design of soft robots. For those interested in the topic of design, there are fantastic options available: one example is *The Science of Soft Robots* edited by Koichi Suzumori, Kenjiro Fukuda, Ryuma Niiyama, and Kohei Nakajima. Another example, more focused on Do-It-Yourself examples is *Soft Robotics: A DIY Introduction to Squishy, Stretchy, and Flexible Robots* by Matthew Borgati and Kari Love. Additionally, given the rapid pace in the field, this book does not seek to capture the latest research examples. Instead, the research examples shown capture the concepts in the clearest and most straightforward way possible.

Aiming to relate to conventional rigid robotics, the book seeks to capture the critical building blocks: sensing, computation, energy storage, and actuation mechanisms. However, at the time of writing this book, the greatest challenge in soft robotics is still actuation. As soft robots aim to reproduce locomotion abilities of living systems, researchers in the field aim to find artificial replicas of natural muscles. As a design target, natural muscles show extreme performance with specific energy up to 40 J/kg and actuation frequency ranging from 0.1 to 300 Hz. Moreover, natural muscles are multifunctional systems, capable to serve as thermal generators (i. e., shivering), springs for impulsive movements, chemical storage, and even have embedded proprioception, or self-sensing.

The structure of the book is as follows: the first chapter introduces soft matter, focusing primarily on elastomers. While foams, gels, colloids, and other soft matter systems have unique and interesting properties, the vast majority of work in Soft Robotics is aimed at elastomer development and processing to produce soft structures. The second chapter describes stretchable electronic conductors, which serve as the basis for deformable resistive and capacitive sensors, as well as building blocks

https://doi.org/10.1515/9783111069418-202

for electro-mechanical actuators, stretchable energy storage, and harvesting systems. The following four chapters describe the four main actuation modalities: fluid powered systems, electro-mechanical transducers, thermally responsive materials, and magnetic actuators. For each chapter, the focus is on how energy is transduced from a specific energy source into a motion that can be used for robotically-relevant mechanical work. The seventh chapter describes adhesion mechanisms, specifically electro-adhesion and gecko-inspired adhesives, which are both relevant for robotics and rely on soft materials for operation. The final chapter summarizes more advanced topics that cover other robotic building blocks, but they are not as established as the previous chapters.

Each chapter includes some fundamental physics, chemistry, or materials science that explains the device operation. For actuator and adhesion chapters, the focus is on the energy transduction mechanism. Different mechanisms are examined in parallel for a deeper understanding. For example, in thermo-mechanical systems the canonical example is shape memory alloys. These metals are not inherently soft, but can be fabricated into deformable structures that contract upon heating. These materials are compared with liquid crystal elastomers, which are soft materials that undergo a structural reorganization when heated. Lastly, a third example is described through phase change materials embedded in soft matrices. These materials typically incorporate a volatile compound that can become a gas at elevated temperature and a stable matrix that prevents the gas from escaping. Although the three systems rely on different phase changes, the fundamental energy transfer mechanism is the same, and the challenges of adding and removing heat rapidly in a robotic component are similar across the systems. Understanding these topics, as well as the overall actuator specific energy and thermo-mechanical energy efficiency, enable a student of Soft Robotics to improve upon the state of the art.

For a bit of historical context, soft robotics is not a new field: patents that are nearly a century old describe soft rubber bladders being used as sphygmomanometers or blood pressure monitoring devices. McKibben air muscles, which are a type of soft actuator and will be discussed at length in Chapter 3, were invented in 1957 per the original patent. Additionally, Figure 1 shows an example from a 1969 patent describing a mechanism for retrieving torpedoes which relies on inflating hydraulic chambers (see the multiple valves at 36, 42, and 62). Knowledge of soft materials doing work as autonomous robotic systems has been established for decades. The recent explosion in popularity of Soft Robotics is due to the ease of prototyping, which enabled wide adoption of fabrication techniques and engaged a broad community of researchers, looking at both fundamentals in the materials, fabrication, and integration, as well as applications across the economy from collaborative manufacturing, to wearables, healthcare and beyond.

This growing community is whom the book is addressed to. The goal of this book is to give the reader the fundamental understanding and practical tools to compare technologies across scales, operation regimes, and fabrication methods. The secondary goal of this book is to teach soft robotics in an interdisciplinary fashion. Soft Robotics now encompasses multiple engineering and scientific disciplines, and a common denomina-

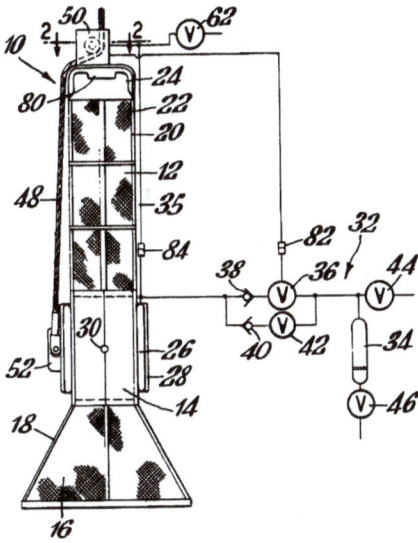

Figure 1: Example of a patent (3647253A) filed in 1969 describing an "Apparatus for retrieving an experimental marine torpedo, or any other load, which is adapted to float on the surface of a body of water, comprising a main inflatable center section and a detachable nose and tail section."

tor is needed to orient a novice and empower them to push the boundaries of the field. I hope the structure of and examples in this book convince others to focus on energy and enable new discoveries and understanding.

In terms of logistics, this book is aimed to aid an instructor teaching a typical semester long course. It can also be pared down to just a quarter course running over 10 weeks. The course can operate purely as lecture-based one, making use of videos from the reference papers to give the student a thorough understanding of the material. However, the best way to teach soft robotics is learning by doing, and therefore the book includes example laboratory sessions, most of which can be completed with food grade materials and inexpensive laboratory equipment. Example problems are given for homework or exam purposes and a project component is strongly encouraged anytime this course is taught. The book is primarily intended for early career graduate students, so first- or second-year PhD or master students, although the content is accessible to well-prepared advanced undergraduates.

Acknowledgment

The breadth of topics covered is a result of productive conversations with colleagues and students across many years. For all their support in putting together the pieces of this book, I would like to thank, in no particular order, Dr. Codrin Tugui, Ang Li, Alexander White, Anatol Gogoj, Dr. Alexander Yin, Dominic Flores, Tirth Thakar, Estefany Toribio-Castillo, Sahib Sandhu, Liam Wilson, Hubert Sliwka, Hao Gu, Katia Ionkin, Dr. Siyoung Lee, Haleh Shahsa, Nour Dowedar, Edward Pomianek, Victor Jimenez-Santiago, and Eva Crowley. The book would not have been possible without the unwavering support of my family, to whom I am forever grateful.

https://doi.org/10.1515/9783111069418-205

Contents

1 Soft matter fundamentals

Contents

Soft materials are the key building blocks for soft machines, and understanding their structure and behavior is critical to design device performance. This chapter focuses on elastomeric materials, which are inherently stretchable and serve as the primary material for soft machine technologies. In parallel, approaches that rely on flexible and foldable materials, such as origami, are evaluated.

1.1 Learning objectives

The main concept to be presented and understood is the mechanical behavior of elastomers, and how it depends on the materials structure and chemistry. Behavior of real-world systems is analyzed and simplified for the context of interest: behavior of elastomers as building blocks of soft machines. Concepts such as toughness, stiffness, and resilience are introduced to better understand the limits of mechanical deformation.

1.2 Background and principles

An *elastomer* is a material capable of returning to its original shape, or close to it, after undergoing large stretch, in one or more dimensions. Most often, elastomers are comprised of long polymeric chains, which are entangled and weakly interacting. As the material undergoes deformation, the chains detangle and align until they reach their maximum length, and the stiffness of the material increases significantly.

    Young's modulus, also known as the *modulus of elasticity*, is a mechanical property of solid materials that measures the strain response in a material to an applied stress, either in tension or compression. In this book, Young's modulus is written as Y and defined as the ratio of stress (σ, with units of pressure, or Pascals) and strain (λ, unitless defined as the ratio of the final length over the initial length, $\lambda = L_{\text{final}}/L_0$). Throughout the book, *stiffness* will be used as an alternative descriptor for Young's modulus, in the range of elastic deformation.

https://doi.org/10.1515/9783111069418-001

$$Y = \frac{\sigma}{\lambda}.$$

(1.1)

For most demonstrations in this book, soft elastomeric materials are chosen to be resilient. *Resilience* is the ability of a material to absorb energy when it is deformed elastically and release that energy upon unloading.

When materials are taken past the elastic limit, they begin to undergo plastic deformation, from which they do not return to the original dimensions, i. e., some permanent deformation occurs. The maximum stress a material can undergo before breaking is called *ultimate strength*, as shown in Figure 1.1. Beyond that point materials typically weaken, meaning the stress required to continue deforming them becomes lower, up to the point of failure. The entire energy applied to the material before failure, visualized as the area under the curve on a stress vs. strain diagram, is shown in Figure 1.2 and is called *toughness*.

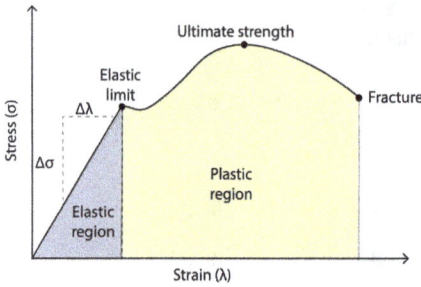

Figure 1.1: Stress vs. strain behavior for a typical elastomer, which undergoes plastic deformation past its elastic limit, leading to fracture.

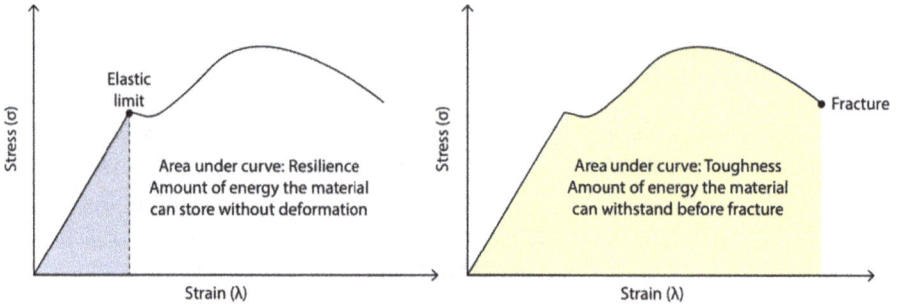

Figure 1.2: Comparison of resilience and toughness: resilience represents the amount of energy a material can withstand without permanent deformation, whereas toughness is the total energy it can absorb up to fracture.

1.3 Mechanical energy in the limits of elasticity

We will discuss various types of soft machines, most, if not all, of which are enabled by the ability of an elastomer to absorb mechanical energy and return to its original shape. The volumetric mechanical energy stored in an elastomer when deformed can be written as

$$e_{mech} = \frac{1}{2}\lambda \times \sigma, \tag{1.2}$$

where
- e_{mech} is the mechanical energy stored, in units of energy per volume,
- λ is the strain in the material, and
- σ is the stress applied to the elastomer, with units of pressure, or N/m^2.

From their definition we know that the stress and strain are proportional to each other, with a factor of the material's Young's modulus (Y, which has units of pressure, or N/m^2). A standard assumption we apply throughout this chapter is that the elastomer is incompressible and has a Poisson's ratio of 0.5. From a simple dimensional analysis it is straightforward to see that the units of volumetric energy density are the same as those of pressure:

$$1\frac{J}{L} = 1\frac{N\,m}{10^{-3}m^3} = 10^3\frac{N}{m^2} = 1\,kPa. \tag{1.3}$$

If we imagine an actuator that can deform a soft elastomer, then we can develop some understanding of how the elastomer properties impact the actuator performance. As a theoretical experiment, we can imagine a soft balloon in which the operator controls the type of elastomer used in the construction of the balloon. The balloon can be constrained so that it does not expand, but just expands against a force sensor. Alternatively, the balloon can expand without any constraints. As shown in Figure 1.3, the total available mechanical energy in an actuator is half the product of the displacement and the blocked force produced. The slope of the line connecting maximum displacement and maximum blocked force is proportional to the material's Young's modulus. Extreme cases such as soft or stiff elastomers are both captured in this view: soft elastomers show large displacement but low force output, whereas the stiffer elastomers have limited displacement but large force outputs.

Writing the energy equation in terms of the stress and strain, we see that the mechanical energy is equal to

$$e_{mech} = \frac{1}{2}\lambda^2 Y. \tag{1.4}$$

The following chapters describe various ways of input energy (e. g., electrical, fluidic, thermal, etc.), which deforms a structure made of elastomeric materials. When the

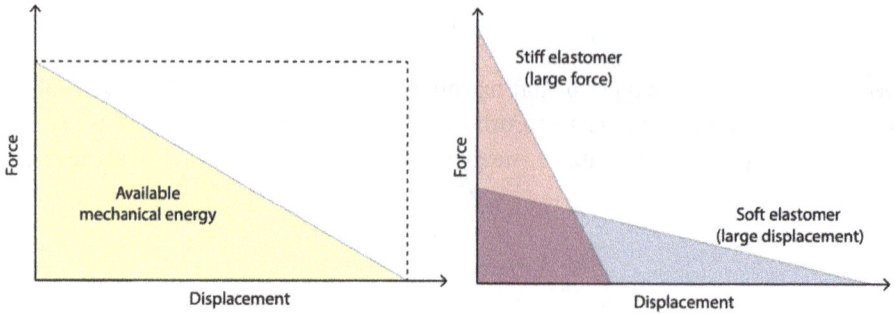

Figure 1.3: Left: trade-off between force in an elastomeric actuator and the displacement in the same system. Area under the curve corresponds to the mechanical energy available. Right: trade-off between using a soft elastomer, which shows more displacement, and a stiffer elastomer, which develops more force.

stress is removed, the material returns to its original shape and dimensions, allowing the soft machine to continue operating. The amount of strain depends on the type of energy input, whereas the Young's modulus is set by the operating conditions and structure of the soft machine.

1.4 Practical considerations: elastomer chemistries

There are many types of elastomeric polymers that can be used in the construction of soft machines. The backbone of the polymer has broad implications on both how the elastomer is made and its properties under the environmental and operating conditions of the soft machine. In this book, we primarily focus on two types of backbones that form siloxanes, carbon–carbon and silicon–oxygen bonds. A prototypical structure for siloxanes is shown in Figure 1.4.

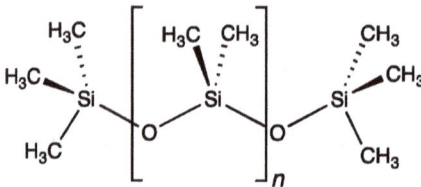

Figure 1.4: Structure of a polydimethylsiloxane molecule with silicon–oxygen bonds along its backbone.

Within the field of soft robotics, the most commonly used siloxanes are known as polydimethlysiloxane (PDMS). These materials have several advantages when being used as polymeric elastomers to build soft machines: they are typically made from inexpensive two-part mixtures that can cure slowly at room temperature or in an accelerated fashion when heat is applied. The materials have good biocompatibility, long calendar life, and low viscoelasticity, so they can quickly respond to deformation. The ease and speed of prototyping with PDMS have been the characteristics that have driven broad adoption for building soft machines.

The are multiple types of elastomers based on carbon backbones: historically, the first elastomeric material used was natural rubber. These materials are generally more difficult to make than silicones because they cannot be produced in from a two-part room temperature curing mixture. However, the different properties promote novel fabrication approaches compatible with specific applications. Figure 1.5 shows the chemical structures of four types of carbon-based elastomers. The specific properties of each depend on the average molecular weight of the polymeric chains, the types of other polymers that are copolymerized, and any side groups included.

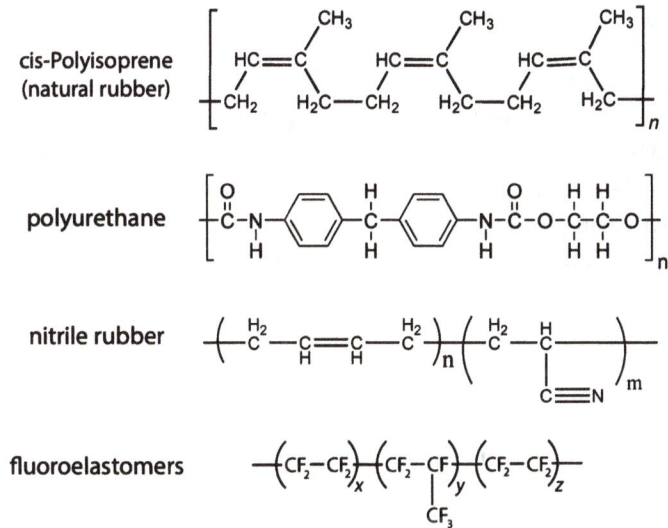

Figure 1.5: Structures of carbon backbone containing polymers, including natural rubber, polyurethane, nitrile rubber, and a fluorinated elastomer.

We outline some baseline principles for selecting carbon-based elastomers:

- *natural rubbers* – polyisoprene material, which typically is used as a thermoset material. The rubber is vulcanized by reaction with sulfur, which forms cross-links between sections of the polymer chain increasing rigidity and durability.
- *polyurethane* – most commonly used as a thermoplastic material, which can be easily deformed when heated and returns to its original properties when cooled. The thermoplastic abilities make polyurethane a suitable candidate for fused filament 3D printing, a type of additive manufacturing.
- *nitrile rubber* – a type of elastomer modified for increase resistance to oils and other common chemicals. Although useful for applications such as laboratory disposable,

nonlatex gloves, the processing required to make nitrile rubber makes it a poor candidate for building soft machines.

– *fluoroelastomers* – typically, the most chemically resistant elastomers, providing increased resistance compared to nitrile rubber for both heat and chemicals. Fluoroelastomers are the material of choice for automotive and aerospace applications where elastomers need to be both deformable and heat resistant.

– *acrylic elastomers* – different from the above materials, because they can be made by curing under ultraviolet light. The reaction typically occurs in the absence of oxygen. Examples will be described when discussing stretchable conductors (e. g., hydrogels) and dielectric elastomer transducers in the following chapters.

1.5 Complex behavior: glass transition temperature, hyperelasticity, viscoelasticity, and hysteresis

All elastomeric materials have something called a *glass transition temperature* T_g, the temperature at which the polymer chains start to move (Figure 1.6). Below T_g, the polymer chains are rigid and have much less mobility. The T_g is influenced by the molecular structure of polymers, molar mass, crystallinity, and thermal history. Lowering the T_g can increase the polymer chains' mobility and consequently increase the conductivity. This is due to the enhanced flexibility of the polymer chains and faster segmental motion.

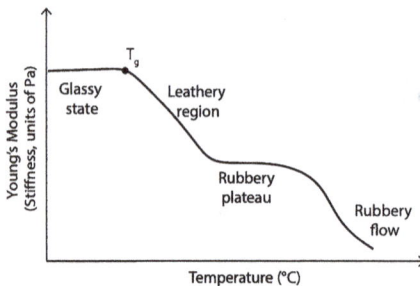

Figure 1.6: Example of the change in material stiffness as a function of temperature, showing the transition from glassy behavior to rubber.

Glass transition temperatures range from $-100\,°C$ for some silicone elastomers to near $-20\,°C$ for fluoroelastomers. The window of heat resistance is similarly shifted, as fluoroelastomers typically exhibit the highest heat resistance, being able to operate up to $250\,°C$. For most soft robotic applications, materials are chosen to maintain flexibility in the range of temperatures comfortable for humans, so PDMS-type materials are most often suitable. Additional advantages for silicones is the ability to operate in low temperatures, below the range where humans lose dexterity, allowing soft machines to perform actions with human-like dexterity.

Hyperelasticity is a term that captures the nonlinear behavior of most real-world elastomers during mechanical deformation. For example, rubber under mechanical strain shows a nonlinear stress response, which is isotropic, meaning that no one direction is different from the other. At the same time the material is incompressible, another term for saying the material has a Poisson's ratio of 0.5. The Poisson's ratio is defined as the ratio of lateral strain over longitudinal strain, and for incompressible materials, the two are proportional. Hyperelastic behavior has the most impact on dielectric elastomer transducers, in particular actuators. Those systems require prestretch to be applied to the elastomer to overcome the initial high stiffness of the material and to reach a plateau in the stress vs. strain curve. Additionally, the strain stiffening behavior of hyperelastic materials helps keep the materials from reaching the breakdown limit during electro-mechanical actuation. More details are presented in the *Dielectric Elastomer Transducers* chapter.

Viscoelasticity describes the ability of a material to respond quickly to an applied deformation. The viscoelastic behavior of a material can be determined through dynamic mechanical analysis. The basic principle is shown in Figure 1.7: an oscillating stress σ is applied to the elastomer, and the responding strain λ is measured. For a purely elastic material, the stress and strain are completely in phase. Meanwhile, for a viscoelastic material, there is a phase lag between the input stress and output strain, labeled as δ. Mathematically, this can be expressed as

$$\sigma = \sigma_0 \sin(\omega t), \tag{1.5}$$
$$\lambda = \lambda_0 \sin(\omega t + \delta), \tag{1.6}$$

where
- $\omega = 2\pi f$, where f is frequency of strain oscillation with units of time,
- t is time, and
- δ is the phase lag between stress and strain.

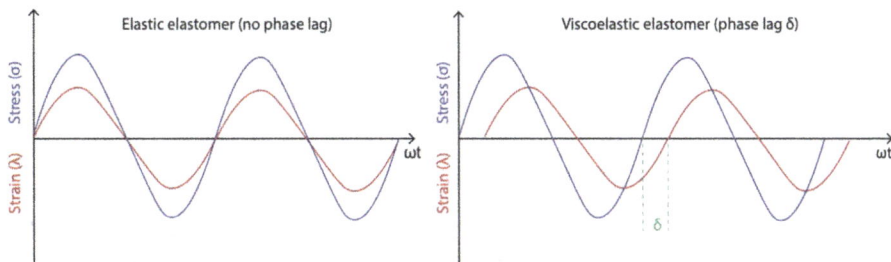

Figure 1.7: Comparison of a purely elastic response without a phase lag between an applied stress (blue) and the response strain (red) with a viscoelastic material with a phase lag δ.

In this notation, we can further define the storage and loss modulus to quantify the relative viscoelastic behavior between materials:

- Storage modulus $E' = \frac{\sigma_0}{\lambda_0} \cos(\delta)$ represents the amount of stored elastic energy in the material;
- Loss modulus $E'' = \frac{\sigma_0}{\lambda_0} \sin(\delta)$ represents the amount of energy dissipated as heat.

The ratio of the loss and storage moduli is a critical parameter, also known as the *loss tangent*, and provides a measure of the damping in the material:

$$\tan(\delta) = \frac{E''}{E'}. \tag{1.7}$$

Elastomers under real loading conditions exhibit a phenomenon called *hysteresis*, the dependence of the state of the elastomer on its deformation history. As shown in Figure 1.8, the energy required to deform a viscoelastic material is slightly larger than that needed to deform its true elastic counterpart. Similarly, when the stress in the material is released, there is less available elastic energy, compared to the purely elastic system. That difference in energy, shown in green in the graph on the right, is energy dissipated as heat by the elastomer. The heat is the result of friction between polymer chains, which are sliding past each other and rearranging when the bulk material is deformed. The amount of heat dissipated depends on the rate of deformation, as slower movement amounts to less friction between the polymer chains.

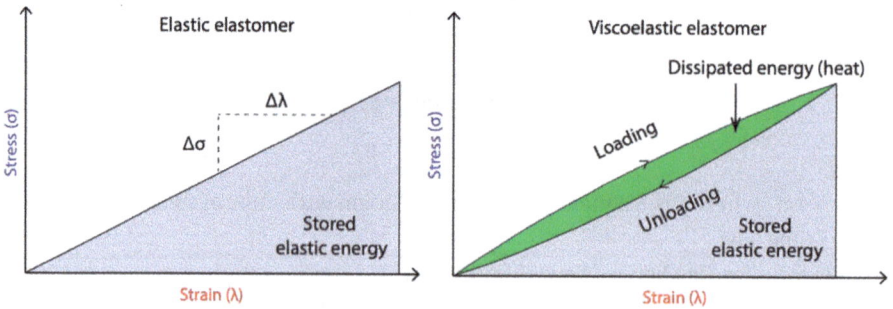

Figure 1.8: Comparison between a true elastomer, where all deformation energy is stored as elastic energy, and a viscoelastic elastomer, which exhibits hysteresis with some energy dissipated as heat.

1.6 Alternative approaches: folding, origami, kirigami

A vast research portfolio exists, which uses flexible materials to build machines that undergo deformation. A particularly rich area of exploration is the use of origami folding techniques to produce structures changing shape reliably. Origami is an effective technique for forming a two-dimensional material, typically paper, into a 3D shape. With roughly 10–12 fundamental folds, the origami approach is capable of creating highly complex structures. A variant of origami, kirigami, is a technique that also incorporates cuts into the 2D material to produce complex 3D shapes. This research space is vast and extremely interesting but conceptually not aligned with the energy-focused approach of this book.

Figure 1.9 shows an example of how origami can be integrated with soft matter to produce programmable deformation. The left side of the figure shows the primitive origami folds needed for building 3D shapes. The right side of the figure shows an origami bellows coated with silicone elastomer, from reference Martinez et al., "Elastomeric origami: programmable paper-elastomer composites as pneumatic actuators". The coating makes the bellow airtight, and during pressurization, the elastomeric material can stretch. The function of the origami is to serve as a deformation guide, producing linear expansion, which is useful for a broad range of robots. There is an advantage of using paper as a strain limiting layer: the paper has a fibrous microstructure, which allows the uncured silicone elastomer to flow around it and cure in place, mechanically interlocking the two structures.

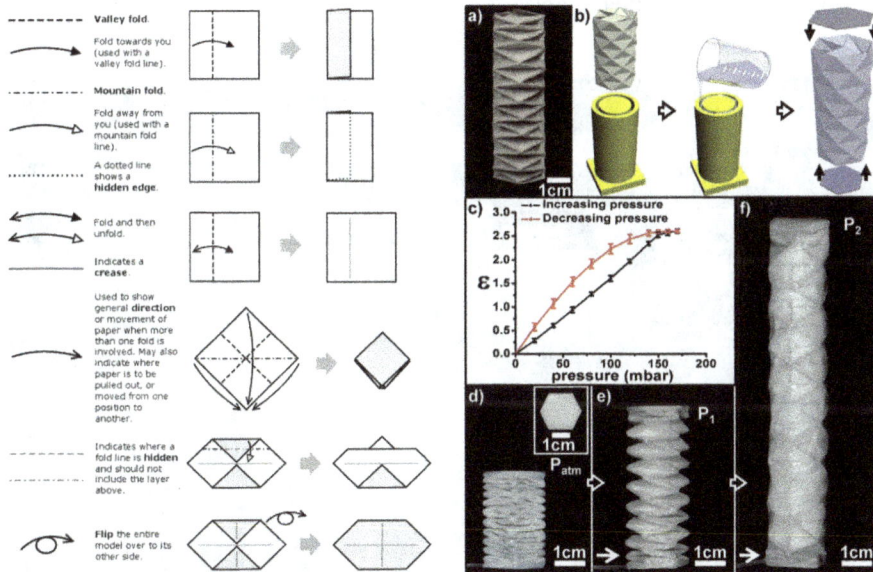

Figure 1.9: From Martinez et al. Left side: instructions on simple origami folds. Right side: research literature example showing a composite paper–silicone bellow that undergoes linear expansion when pressurized. Reproduced with permission.

1.7 Example laboratory session

In this laboratory session the students prepare elastomer samples (e. g., Ecoflex 00-30, Ecoflex 00-50, and Dragon Skin 20 from SmoothOn) as thin films. The next step will be to test the mechanical properties of the elastomers in tension. An alternative material is an acrylic elastomer 3M VHB 4910, which can be purchased directly as a cured film.

Note: The listed silcone elastomers will require curing in an oven at temperatures above 70 textdegree C. If an oven is not available, a fast curing material, such as Eco-Flex 00-35 FAST is highly recommended.

Figure 1.10 shows all the materials needed, including
- elastomer part A,
- elastomer part B,
- laboratory scale,
- cutting mat,
- laboratory stand,
- mixing cups with wooden stirrers,
- calipers to accurately measure sample dimensions,
- set of hooked laboratory weights for tensile testing,
- array of clips and rulers to measure displacement,
- mold for curing, and
- dog bone die for cutting samples.

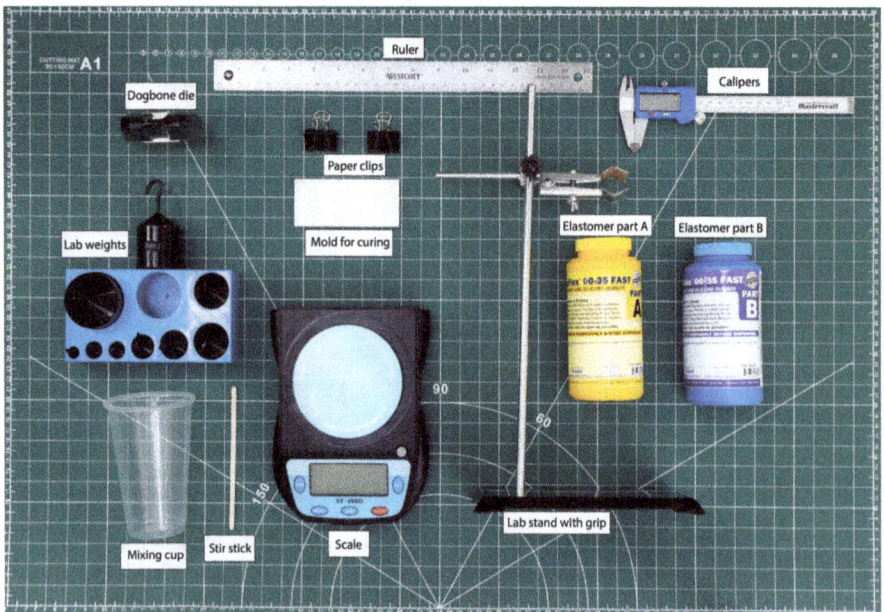

Figure 1.10: Preparation for laboratory example 1: all of the materials required to test elastomeric materials in tension to determine Young's modulus.

Figure 1.11 shows the first four steps in the elastomer making process. The students need to mix equal amounts (by weight) of parts A and B of each elastomer type in a mixing

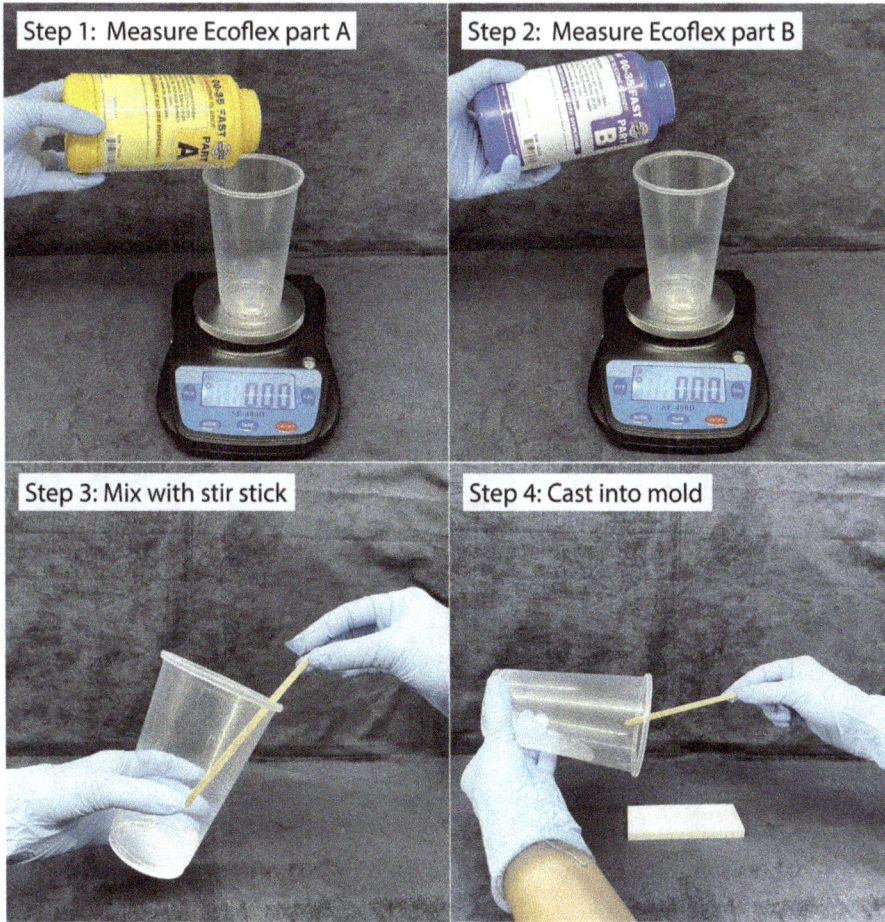

| Step 1: Measure Ecoflex part A | Step 2: Measure Ecoflex part B |
| Step 3: Mix with stir stick | Step 4: Cast into mold |

Figure 1.11: Steps 1–4 in testing elastomer stiffness.

cup. To do this, weigh out an empty mixing cup on the laboratory scale and tare the scale. Then transfer ≈ 20 grams of part A to the mixing cup (**Step 1**). Record the exact weight you transferred. Tare the scale again. Then transfer the same mass of part B into the same mixing cup (**Step 2**). Record the final weight of part B. If the amount of B is off by more than 5 % than that of A, then add more A to compensate. Then use a wooden stirrer to mix both components for 30 seconds (**Step 3**). Lastly, pour out the mixture into a flat mold (**Step 4**) and transfer it to an oven, set to 70 °C. Start a 15 minute timer, then at the end of it remove the cured sample using oven gloves. Peel the elastomer from the mold (**Step 5**) and proceed.

To measure the Young's modulus, students need to be able to apply stress and measure strain on an elastomer sample. Figure 1.12 shows the next stpes in this sequence: starting with cutting dog bone shaped samples using the available die cutters. For ten-

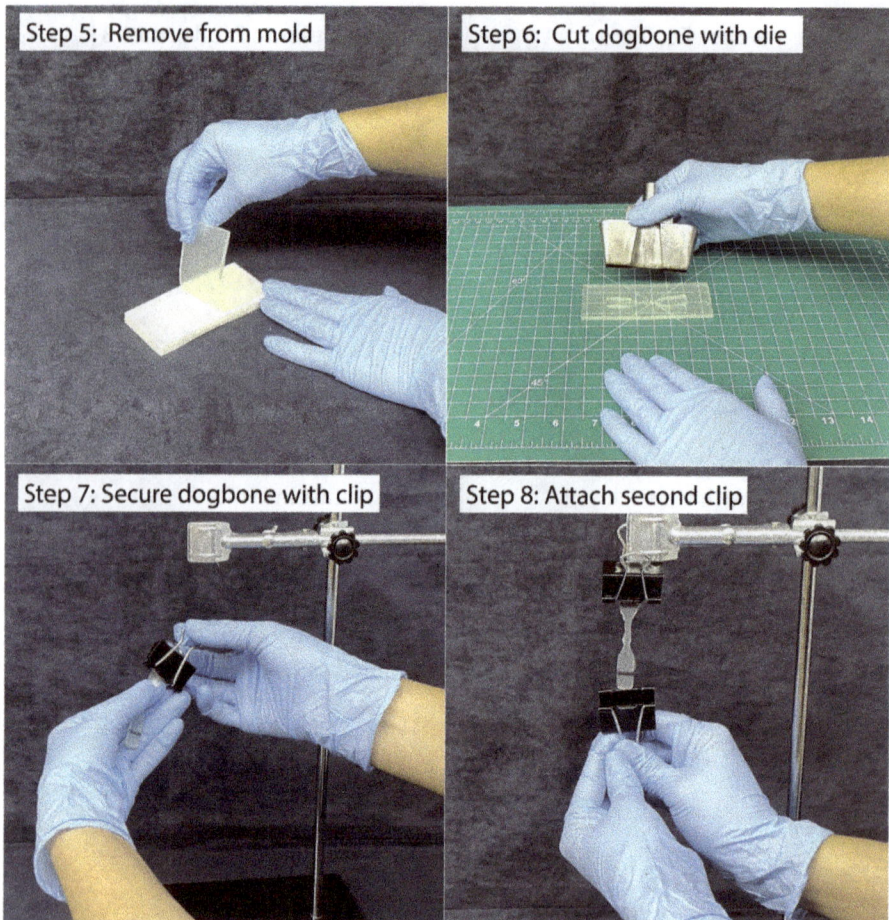

Figure 1.12: Steps 5–8 in testing elastomer stiffness.

sile testing, the samples are aligned vertically and attached to a stand. The last steps are shown in Figure 1.13: weights are added to the bottom of the sample, causing it to elongate. The mass of the weights and the sample dimensions determine the stress on the elastomer. The elongation is a measure of strain therefore, by plotting stress vs. strain students will be able to find an approximate value for Young's modulus.

1. Cut dog bone samples from each elastomer (**Step 6**). Make five samples for each elastomer type, including the three silicone elastomers cured and VHB.
2. Secure the top of the sample with clips on a vertical stand (**Step 7**).
3. Attach clips at the bottom to allow you to hook weights onto the clips and elongate the sample (**Step 8**).
4. Make two marks on the dog bone and record the distance between them (**Step 9**), that is, your initial length L_0.

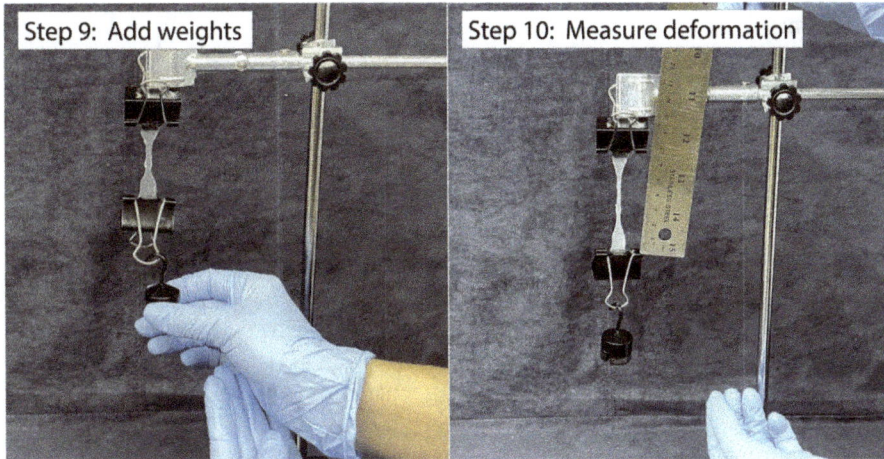

Figure 1.13: Steps 9 and 10 in testing elastomer stiffness.

5. Begin to add weights and record the deformed length L (**Step 10**). Use a graphing software to record weight vs. length in a table for each sample.
6. Using sample dimensions, convert weight and length into stress (in kPa) and strain (in %) and plot them relative to each other.
7. Using the slope of the curve, determine the average and standard deviation for the Young's modulus of each of the four elastomers.
8. In a table, record results and compare them with the predicted values from the prelab report.

1.8 Example problems

Problem 1. How much mechanical energy is stored in an elastomer with a Young's modulus stretched uniaxially 100 % relative to its original size? Assume that the elastomer is a 10-cm-long strip, with width 1 cm and thickness 1 mm. Report your result in Joules.

Solution. The mechanical energy density (e_{mech}) for a uniaxial stretch is given by the formula

$$e_{mech} = \frac{1}{2}\lambda^2 Y, \tag{1.4}$$

where
- e_{mech} is the mechanical energy density (J/m^3),
- λ is the stretch ratio, defined as $\lambda = \frac{L_{final}}{L_{initial}}$, and
- Y is Young's modulus (Pa).

Given:
- $\lambda = 2$ (100 % relative stretch),
- $Y = 1\,\text{MPa} = 10^6\,\text{Pa}$,
- dimensions: $L_0 = 0.1\,\text{m}$, $w = 0.01\,\text{m}$, $t = 0.001\,\text{m}$.

The volume of the elastomer is

$$V = L_0 \cdot w \cdot t = 0.1 \cdot 0.01 \cdot 0.001 = 10^{-6}\,\text{m}^3.$$

The total mechanical energy U is obtained by multiplying e_{mech} by the volume:

$$U = e_{mech} \cdot V.$$

Substituting the values, we get

$$e_{mech} = \frac{1}{2}\lambda^2 Y = \frac{1}{2}(2)^2(10^6) = 2 \times 10^6\,\text{J/m}^3,$$
$$U = (2 \times 10^6) \cdot (10^{-6}) = 2\,\text{J}.$$

Answer: The total mechanical energy stored in the elastomer is 2 J.

Problem 2. A thin strip of elastomer (thickness t_e) is stretched by a uniform strain λ and then bonded to a thin, flexible polyester substrate (thickness t_p) that is flexible but not stretchable. After bonding, the system is released and the bilayer bends into an arc to relieve the mismatch in strain. This assembly process is used frequently in soft robotics. Assume the following:
- The elastomer behaves as a linear elastic material with Young's modulus Y_e.
- The polyester is also linear elastic with Young's modulus Y_p, but it does not undergo any initial strain.
- The width of both layers is much greater than their thickness, and plane strain conditions apply.
- The bilayer bends into a circular arc with radius of curvature R.

Determine:
a   Derive an expression for the radius of curvature R of the bilayer strip as a function of the applied strain ϵ, the moduli Y_e, Y_p, and the layer thicknesses t_e, t_p.
b   Use the following simplified formula to estimate the curvature:

$$\frac{1}{R} = \frac{6\lambda}{t} \cdot \frac{Y_e t_e^2}{Y_e t_e^2 + Y_p t_p^2} \tag{1.8}$$

where $t = t_e + t_p$ is the total thickness of the bilayer.
Given values:
- $\epsilon = 0.1$,

- $t_e = 1\,\text{mm}$,
- $t_p = 0.5\,\text{mm}$,
- $Y_e = 0.5\,\text{MPa}$,
- $Y_p = 2\,\text{GPa}$

Calculate the approximate radius of curvature R.

c   Plot the radius of curvature as a function of the thickness of the polyester layer.

Problem 3. A soft elastic material is formed using a two-stage interpenetrating polymer network (IPN) strategy:
- In the first stage, a polymer network (Polymer A) with a Young's modulus of 500 kPa is mechanically stretched to 100 % uniaxial strain (i. e., stretched to double its original length).
- While held at this stretched configuration, a second monomer is introduced and allowed to diffuse into the open space of Polymer A. Once the diffusion is complete, Polymer B is cured in place, forming an interpenetrating network. Polymer B has a Young's modulus of 1 MPa.
- After curing, the stretch is released, and the material is allowed to return toward its equilibrium configuration.

Assume the following:
- Both polymers behave as linearly elastic materials for small deformations.
- The IPN is perfectly bonded, and the final composite seeks a mechanical equilibrium where the internal stresses in A and B balance.
- There is no slip or relative motion between the networks.
- All deformation is uniaxial and along the same axis.

Determine answers to the following questions:
1. After the release of the external load, what is the final strain (λ) in the IPN at mechanical equilibrium?
2. What is the internal stress (σ) in each polymer (A and B) at equilibrium?
3. What is the effective Young's modulus of the IPN in this direction, assuming small additional uniaxial deformations?

Bibliography

Ramses V Martinez et al. "Elastomeric origami: programmable paper-elastomer composites as pneumatic actuators". In: *Advanced functional materials* 22.7 (2012), pp. 1376–1384.

2 Stretchable conductors

Contents

Stretchable conductors are the pathway to delivering electrical energy into soft machines and for getting electrical signals out. This chapter focuses on stretchable electrical circuit building blocks, in particular resistors and capacitors, which can be used in sensors, actuators, tunable adhesives, energy storage, and harvesting.

2.1 Learning objectives

The main concept to be presented and understood is the relationship between structure and properties in electrical components, which can be deformed and retain electrical conductivity. Three mechanisms for electrical conductivity are explained: metallic, ionic, and delocalized π orbitals, particularly in carbon allotropes. Trade-offs between conductivity, density, and behavior at large strains are presented in the context of soft robotic applications.

2.2 Background and principles

Electrical components offer different responses to the flow of electrons through the circuit. There are two parameters used to control and quantify the movement of electrons, voltage and current. *Voltage*, which is also known as (electrical) potential difference, electric pressure, or electric tension is the difference in electric potential between two points and has units of *Volts*. In autonomous systems, voltage is provided typically by a battery, an electro-chemical cell, which undergoes reversible reactions that produce an output electrical bias. *Current* is the flow of charged particles, such as electrons or ions, moving through an electrical conductor or space. It is defined as the net rate of flow of electric charge through a surface and has units of *Amperes*, also equal to Coulombs per second, where Coulomb is a unit for charge.

Depending on their response to an applied voltage, we consider three types of components as the fundamental building blocks, as shown in Figure 2.1. In electronic circuits,

https://doi.org/10.1515/9783111069418-002

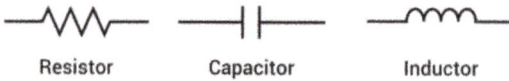

Resistor   Capacitor   Inductor

Figure 2.1: Symbols used in electric circuit diagram for resistors, capacitors, and inductors.

resistors offer some degree of opposition to flow of electrons when a specific bias voltage is applied. These components are used to reduce current flow, adjust signal levels, to divide voltages, bias active elements, and terminate transmission lines, among other uses. Fixed resistors have resistances that only change slightly with temperature, time, or operating voltage. Variable resistors can be used to adjust circuit elements (such as a volume control or a lamp dimmer) or as sensing devices for heat, light, humidity, force, or chemical activity.

A *capacitor* consists of two conductors separated by a nonconductive region. The nonconductive region can either be a vacuum or an electrical insulator material known as a dielectric. A parallel plate capacitor can only store a finite amount of energy before dielectric breakdown occurs. The capacitor's dielectric material has a dielectric strength U_d (in units of Volts per meter) which sets the capacitor's breakdown voltage at $V_{bd} = d \times U_d$, where d is the thickness of the dielectric.

Though not as widely used in building soft machines as capacitors and resistors, some applications exist for soft *inductors*. These components, also known as coils, chokes, or reactors, are passive two-terminal electrical components that store energy in a magnetic field when an electric current flows through it. An inductor typically consists of an insulated wire wound into a coil.

2.3 Combinations of resistors and capacitors

Often in the development of soft machines, it is necessary to combine resistors and capacitors to produce different responses in the voltage and current in the system. The behavior of inductors in relationship with resistors and capacitors is left to be discussed in the chapter focused on *Magnetic Soft Machines*.

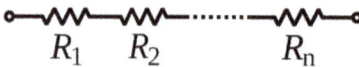

$$R_1 \quad R_2 \qquad R_n$$

Figure 2.2: Diagram showing how resistors are added in series to increase the opposition to current flow.

The equivalent resistance of a circuit in which n resistors of resistance $R_1, R_2, \ldots, R_n$ are added in series is equal to

$$R_{eq} = R_1 + R_2 + \cdots + R_n. \tag{2.1}$$

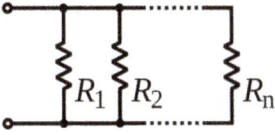

Figure 2.3: Diagram showing how resistors are added in parallel.

The equivalent resistance of a circuit in which n resistors of resistance $R_1, R_2, \ldots, R_n$ are added in parallel is equal to

$$R_{eq} = \frac{1}{R_1} + \frac{1}{R_2} + \cdots + \frac{1}{R_n}. \tag{2.2}$$

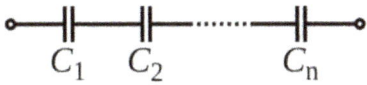

Figure 2.4: Diagram showing how capacitors are added in series.

The equivalent capacitance of a circuit in which n capacitors of resistance $C_1, C_2, \ldots, C_n$ are added in series is equal to

$$C_{eq} = C_1 + C_2 + \cdots + C_n. \tag{2.3}$$

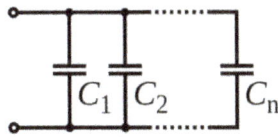

Figure 2.5: Diagram showing how capacitors are added in parallel.

The equivalent capacitance of a circuit in which n capacitors of resistance $C_1, C_2, \ldots, C_n$ are added in parallel is equal to

$$C_{eq} = \frac{1}{C_1} + \frac{1}{C_2} + \cdots + \frac{1}{C_n}. \tag{2.4}$$

The simplest combination of a resistor and a capacitor is called an *RC circuit* and consists of a resistor and a charged capacitor connected to one another in a single loop, without an external voltage source. Once the circuit is closed, the capacitor begins to discharge its stored energy through the resistor. The voltage across the capacitor, which is time-dependent, can be found by using Kirchhoff's current law. The current through the resistor must be equal in magnitude (but opposite in sign) to the time derivative of the accumulated charge on the capacitor.

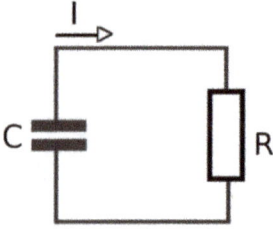

Figure 2.6: Diagram showing the simplest circuit combining a resistor and a capacitor.

Mathematically, this can be written as

$$C\frac{dV}{dt} + \frac{V}{R} = 0.$$
(2.5)

Solving this equation for V yields the formula for exponential decay:

$$V(t) = V_0 e^{-\frac{t}{RC}}$$
(2.6)

where V_0 is the capacitor voltage at time $t = 0$. Often, it is important to know how fast voltage drops in an RC circuit. The time required for the voltage to fall to $\frac{V_0}{e}$ is called the RC time constant and is given by

$$\tau_{RC} = R \times C.$$
(2.7)

2.4 Energy considerations

The behavior of electronic circuits relevant to soft machines is best understood in the context of energy flow through the system. For resistors, their behavior is controlled by *Ohm's law*, which shows the following relationship between voltage V, current I, and resistance R:

$$V = R \times I.$$
(2.8)

In dimensional analysis, the units of voltage and current can be multiplied to give units of power: 1 Volt $\times$ 1 Ampere = 1 Watt. Accordingly, the power dissipation in a resistor can be written as

$$P_{resistor} = V \times I = \frac{V^2}{R} = I \times R^2.$$
(2.9)

The dissipated energy in the resistor is then equal to the product of time t and power dissipation:

$$E_{resistor} = P_{resistor} \times t.$$
(2.10)

A parallel plate capacitor is characterized by its capacitance, which is written as

$$C = \epsilon_0 \epsilon_r \frac{A}{d},$$ (2.11)

where
- C is the capacitance of the capacitor,
- ϵ_0 is the vacuum permittivity, equal to 8.85×10^{-12} F/m,
- ϵ_r is the dielectric constant of the material, a unitless number,
- A is the area of the dielectric plate, and
- d is the thickness of the dielectric.

Meanwhile in a capacitor, the amount of energy stored is written as

$$E_{capacitor} = \frac{1}{2}CV^2 = \frac{1}{2}QV.$$ (2.12)

Given this fundamental knowledge, the rest of the chapter is devoted to understanding how real-world materials can be made into structures that, when deformed, retain electrical conductivity. The parameters introduced above, capacitance, resistance, and geometric dimensions, can all vary due to mechanical deformation, as will be explained below.

2.5 Fundamental origins of electrical conductivity

For a material to be *electrically conductive*, the electrons or ions must be free to move. Within atoms, electrons occupy characteristic energy levels, as dictated by quantum mechanics. As shown in Figure 2.7, if the Fermi level falls within the band that electrons can occupy, then the electrons can easily move and allow the material to conduct electricity. In practice, two types of electrical conductors are relevant to soft machines: metals, which have the highest conductivities on average, and carbon allotropes, different arrangements of atoms, including graphene and carbon nanotubes. Similarly, ionic conductors are widely adopted and rely on conduction due to atomic species (ions) traveling, each carrying an electrical charge. The resistivity of ionic solutions, also known as electrolytes, varies tremendously with concentration and temperature.

It is not sufficient for a material to be electrically conductive to be useful in a soft machine. The conductive material needs to be incorporated within the broader structure and stay electrically conductive when the structure is deformed. Under this constraint, two main types of conductor emerge:

1. *Liquid conductors:* liquid metals, in particular eGaIn, a eutectic mixture of gallium and indium, and electrolytes, using water, organic compounds, or ionic liquids as the solvent, with dissolved species serving as the charge carriers. These conductors need to be encapsulated to prevent the liquid from leaking out of the structure.

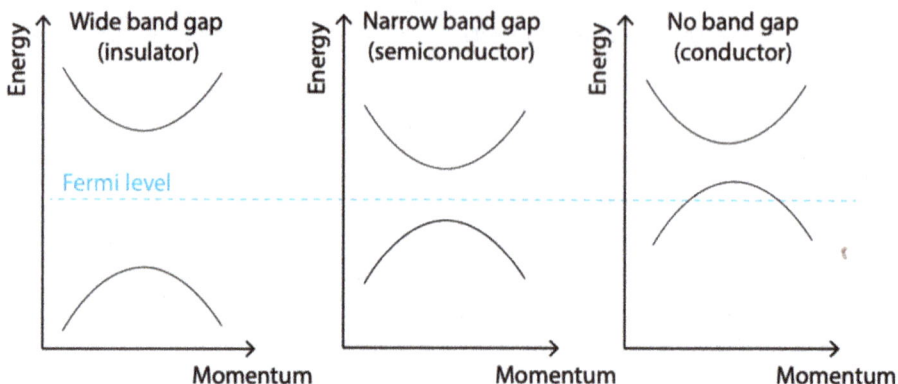

Figure 2.7: Comparison of the band gap in insulators (large gap), semiconductors (small gap), and conductors (no gap).

2. *Deformable solids:* conductors, either carbon, metal, or ceramic-based, arranged in a configuration that allows electrons to flow in different deformation states. Typically, conductivity is due to ballistic electron transport between particles and transport within band gaps within the particles. These conductors need to be supported by an elastomeric substrate that returns them to the original shape after deformation.

The trade-off between the two approaches emerge from the different states present: liquids maintain a conductive path at all times through the liquid but need to be encapsulated to prevent leakage or evaporation of volatile components. Solid conductors are composites, where a conductive particle is embedded in a stretchable elastomeric matrix. The conductive particles typically increase the stiffness of the composite, relative to the pure elastomer. When the conductive particles are sufficiently close and a current is applied across the composite, electrons travel two different ways: via ballistic transport between particles and across delocalized π orbitals within the particles. The advantages and disadvantages of each type shown in Figure 2.8 can be discussed relative to each other:

- *Hydrogels* contain a conductive electrolyte, which forms a gel with a polymer network dissolved in the liquid. The polymerization reaction can be started with ultraviolet (UV) light, allowing for rapid fabrication on demand. The elastic behavior of the hydrogel is due to the polymer, and the properties of the gel depend on the hydration state of the system. An encapsulating layer is typically needed to prevent the dehydration of hydrogels.
- *Ionogels* are a variant of hydrogels, where the water is replaced by an ionic liquid. The strong intermolecular forces between the anion and cation in the ionic liquid prevent evaporation, but ionogels typically have lower conductivities compared to hydrogels.

Hydrogel conductor

Encapsulating layer

Liquid electrolyte

Ions dissolved
in electrolyte

Gel forming
polymer chain

Mobile ions
in electrolyte

Carbon particle conductor

Ballistic e
transport

Delocalized
π electrons

Elastomeric matrix

Carbon particle

Liquid metal conductor

Electrons in metal
atom valence band

Encapsulating layer

Liquid metal

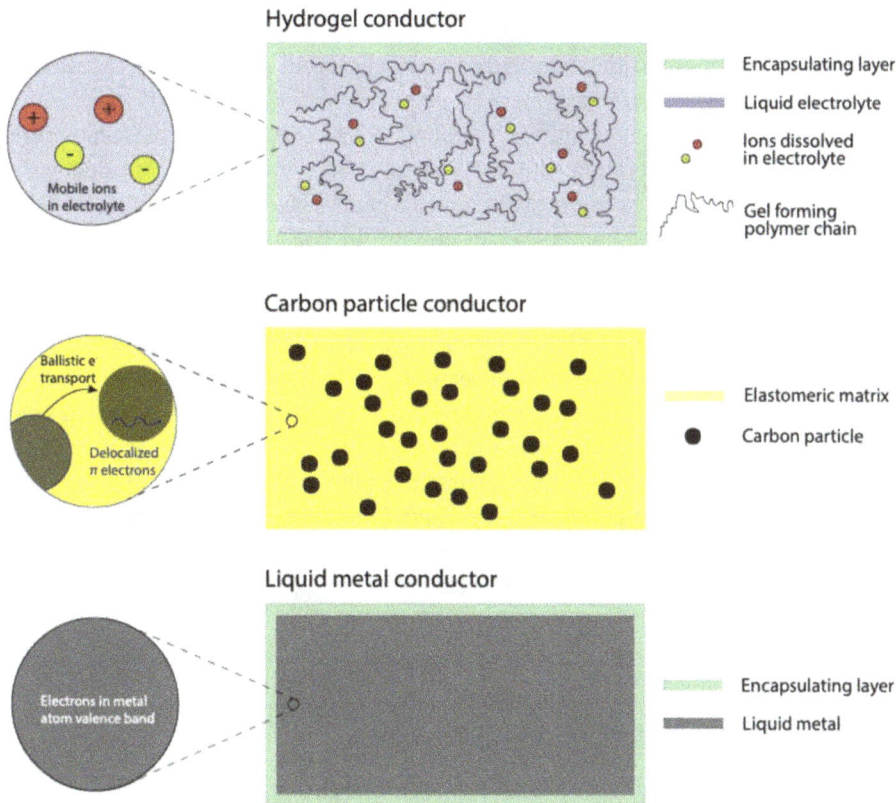

Figure 2.8: Detailed structures of common stretchable conductors, including hydrogels, carbon-elastomer composites, and encapsulated liquid metals. The mechanism for conduction is included in an inset for each type.

– *Carbon-based composites* are solid materials, which have lower conductivities compared to metals, but can withstand large deformations and do not need to be encapsulated. A modification of this approach uses high-aspect ratio conductors, such as *carbon nanotubes* (CNTs), which allow for larger deformation before percolative conductivity is interrupted. Yet another variant uses metallic nanoparticles, such as *silver nanowires* (AgNWs), which have similar behavior to CNTs. One drawback of AgNWs is their reactivity to oxygen, which causes the conductors to tarnish over time and lose conductivity. For all these composites, increasing the conductivity within the particle is not sufficient. The interface between the particle and elastomer must also be tailored to have minimal resistance to ballistic transport.

– *Liquid conductors*, most often metals, have the highest conductivities available, on par with solid metals. A drawback of liquid metals is the high density, in the 6–12 grams per milliliter range, so nearly an order of magnitude higher than water and ionic liquids. Another drawback is the need for encapsulation to prevent leakage of the conductor. In contrast, hydrogels use an internal polymer network, which gives structure to the liquid and reduces the need for encapsulation to provide a restoring force for deformation. One of the most popular materials is the *liquid metal eGain*, an eutectic mixture of gallium and indium, which is liquid at room temperature. A challenge with operating with eGaIn is the formation of a Gallium oxide at the surface of the metal in contact with air. This oxide reduces processibility of eGaIn and requires tailored approaches to produce high-aspect ratio shapes of liquid metal.

With the above conductors available, the behavior of stretchable resistors and capacitors can be evaluated from a geometric perspective. The specific response to deformation for a resistor depends on the conductive path, the change in cross-sectional area, and other factors. Typical approaches aim to achieve a linear change in resistance ($\frac{\Delta R}{R_0}$) in response to a change in strain for usability in applications such as strain gauges. For a capacitor, the change in capacitance is straightforward to model. Assuming that the material is incompressible, the increase in area causes a proportional decrease in the thickness of the dielectric to keep volume constant, or $\lambda_{\text{area}} = \lambda_{\text{thickness}}$:

$$V = A_0 d_0 = Ad = \lambda_{\text{area}} A_0 \frac{d_0}{\lambda_{\text{thickness}}}. \tag{2.13}$$

For the capacitance, the change can be written as

$$C_0 = \epsilon_0 \epsilon_r \frac{A_0}{d_0}, \tag{2.14}$$

$$C_{\text{final}} = \epsilon_0 \epsilon_r \frac{A}{d} = \epsilon_0 \epsilon_r \frac{\lambda_{\text{area}} A_0}{\frac{d_0}{\lambda_{\text{thickness}}}} = \lambda_{\text{area}}^2 C_0. \tag{2.15}$$

2.6 Alternative approaches: corrugated metals and conductive polymers

There are two variants of stretchable conductors that offer some unique advantages over the materials listed above. One approach is applying a layer of metal when the elastomer is mechanically prestretched to a maximum limit for the expected application. In this stretched state, an ultrathin metallic layer is deposited, most commonly via sputter coating. When the prestretched elastomer is released and seeks to return to its original shape, the metallic layer resists compression. The composite accommodates strain by forming a corrugated surface. The steps of the process are shown in Figure 2.9.

Conductive polymers have been explored for decades, since it was established that polyacetylene can conduct electricity along its backbone via electrons in delocalized π orbitals. Polyacetylene is difficult to process, so alternative materials have been developed, including PEDOT:PSS, the structure of which is shown in Figure 2.10.

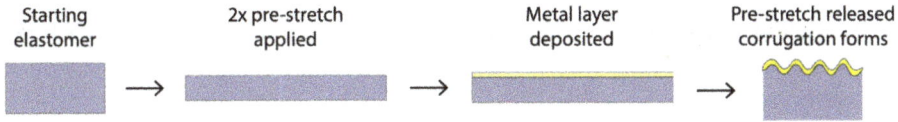

Figure 2.9: Left to right, a schematic of the steps an elastomer undergoes to be coated with metal when fully stretched. When the applied strain is released, a corrugated pattern is formed to accommodate the difference in strain.

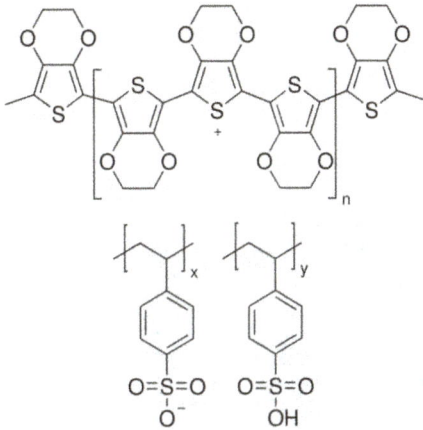

Figure 2.10: Structures of the two main components of the conductive polymer PEDOT:PSS, conductive component on top (PEDOT) and processing additive below (PSS).

The polymer is a two-component system, in which PEDOT (Poly(3,4-ethylenedioxythiophene)) provides electrical conductivity, whereas PSS (polystyrene sulfonate) improves processibility. Specifically, the PSS acts as a counterion to balance the charge and improve the water solubility, as the PEDOT carries positive charges, and the PSS is negatively charged, together forming a macromolecular salt. For the purpose of building soft machines, conductive polymers are similar to carbon particles, exhibiting both ballistic conduction between polymer domains and conduction via π orbitals within the polymer domain. However, the ability to blend polymers, produce interpenetration networks, or form copolymers greatly enhances the processibility of the conductive polymers and allows for fine tuning of the interface between the polymer and the supporting elastomeric matrix.

2.7 Example laboratory session

In this laboratory session, students will prepare stretchable resistors and then test the electrical properties of the devices in response to mechanical deformation. The resistors utilize Eco-Flex 00-35 FAST as the baseline material and a liquid metallic eutectic mix-

ture of gallium and indium (also known as eGaIn). The materials we will need are listed below and shown in Figure 2.11:
- cutting mat,
- elastomer part A,
- elastomer part B,
- scale,
- mixing cups and stir sticks,
- two-part molds: one with a channel and one flat,
- syringes and needles,
- connectors,
- multimeter to measure resistance, and
- liquid metal.

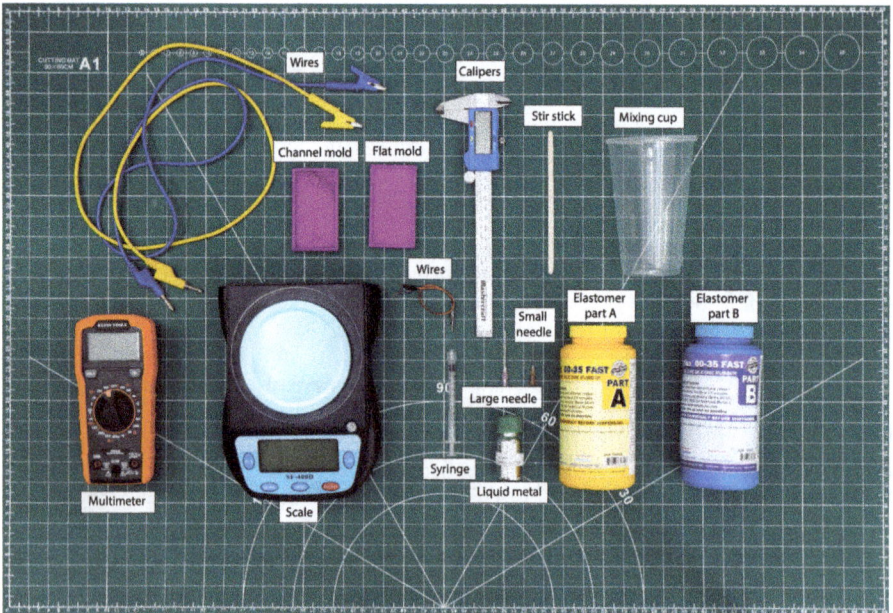

Figure 2.11: Preparation for laboratory example 2: all of the materials required to test stretchable resistors with liquid metal as the conductive material.

To make resistors, students will need to cast two part molds, which contain channels to be filed with a conductive substance. Mix and cure elastomer to cast the channel first (**Steps 1–4**). After the channel part is cured (**Step 5**), start curing elastomer in the flat mold (**Step 6**). Allow the material to cure for approximately 1 minute and 30 seconds, then transfer the channel part onto the flat piece as shown in Figure 2.13 (**Step 7**). Transferring too early causes the channel to fill with uncured elastomer. Transferring

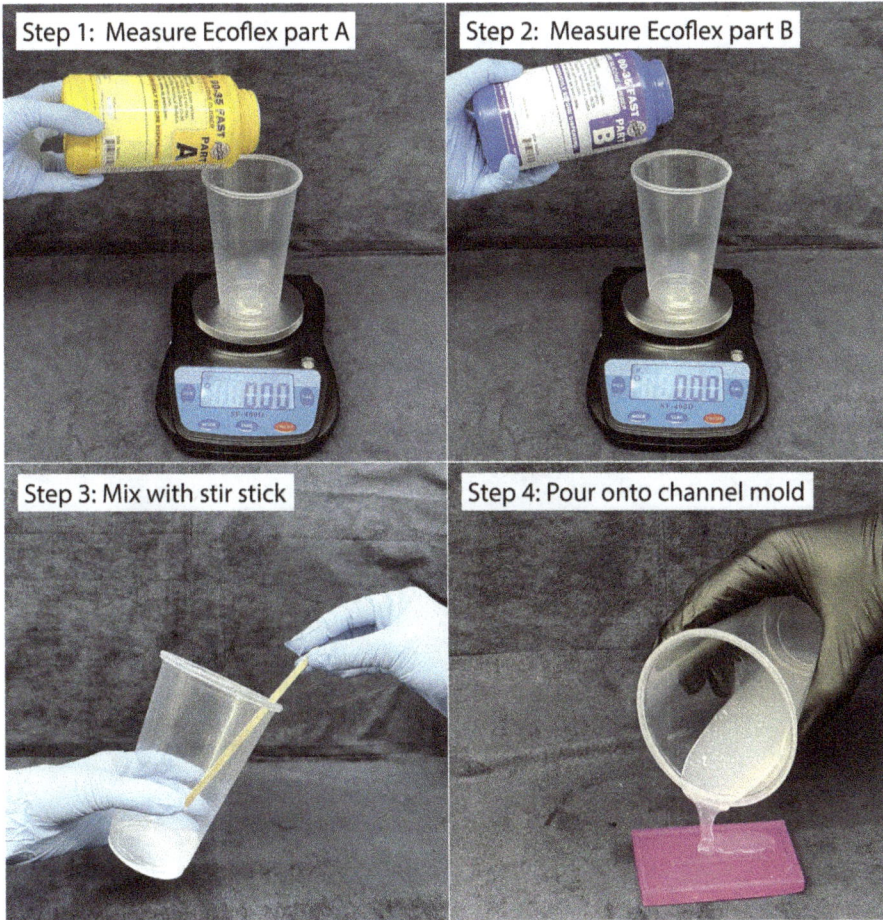

Figure 2.12: Steps 1 through 4 in assembling stretchable resistors.

too late causes the two parts to not bond. When the entire assembly is cured, remove it from the mold (**Step 8**).

Then use syringes to inject the conductor into the channel, as shown in Figure 2.14 (**Step 9**). Insert needles into each reservoir and leave one needle open to lab atmosphere to eject the air. Fill a syringe with the conductive material and inject through the larger one of the needles (**Step 10**). Once the channel is filled, remove the needles and replace them with pin connectors. Add more uncured Ecoflex 00-35 around the insertion point and allow the device to cure for 5 minutes to ensure the connectors are secure. Deform the device as demonstrated in Figure 2.14 and record its resistance (**Steps 11–12**). Compare channels of different sizes and profiles as available.

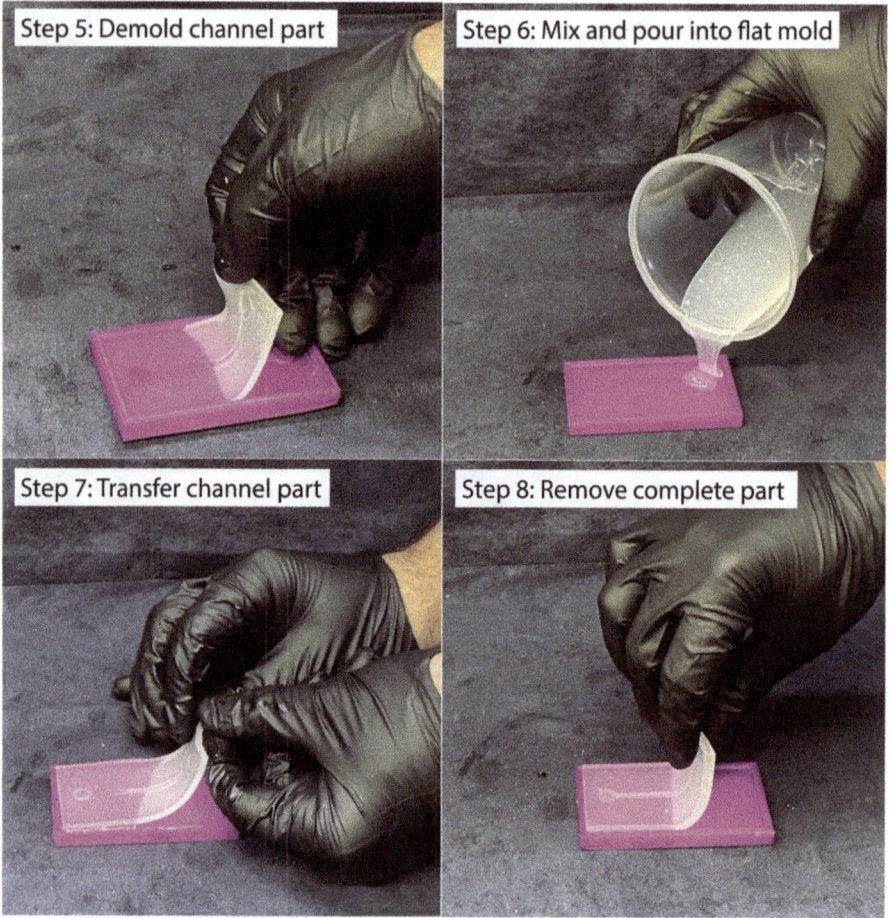

Figure 2.13: Steps 5–8 in assembling stretchable resistors.

2.8 Example problems

Problem 1. Find an equation to estimate the resistance of a cylindrical trace of liquid metal (eGaIn) changes with deformation via stretching due to changes in its length and cross-sectional area.

Solution. The resistance R of a conductor is given by

$$R = \rho \frac{L}{A},$$

where R is the resistance, ρ is the resistivity of the material (in Ωm), L is the length of the conductor, and A is the cross-sectional area.

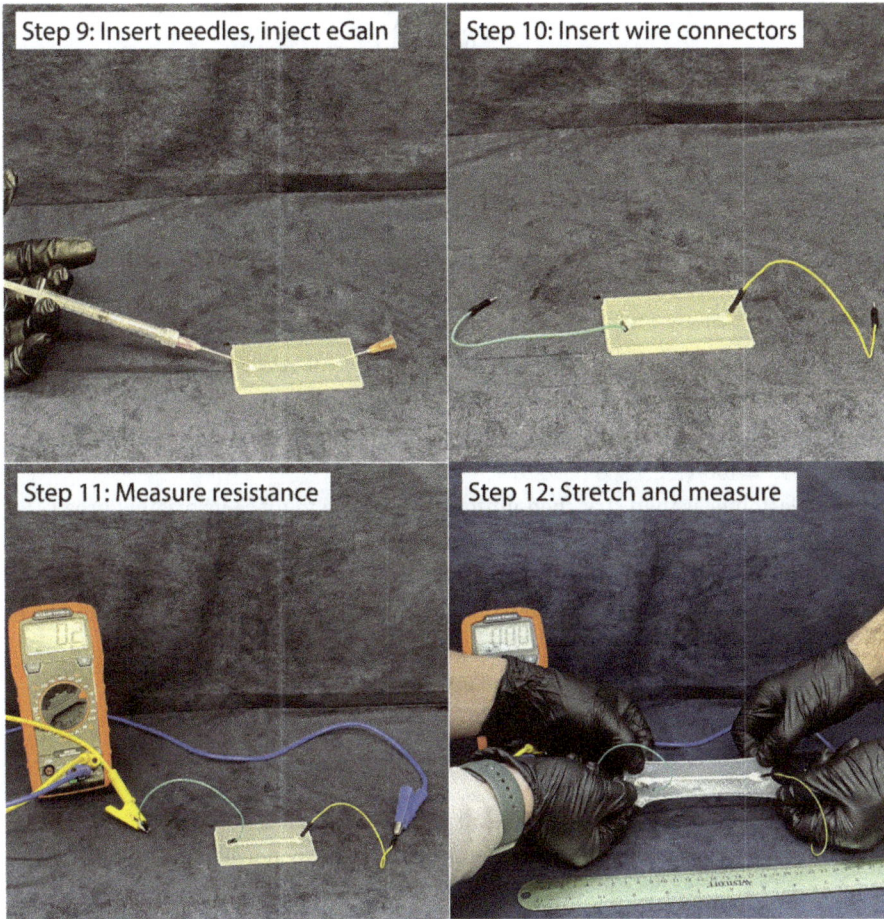

Figure 2.14: Steps 9–12 in assembling stretchable resistors.

Effects of Stretching: a. Increase in length L: when the liquid metal trace is stretched, its length increases. Since resistance is directly proportional to length, this contributes to an increase in resistance.

b. Decrease in cross-sectional area A: stretching also reduces the cross-sectional area because the material has a Poisson's ratio of 0.5 and is incompressible:

$$AL = A_0 L_0,$$

where A_0 and L_0 are the initial area and length, respectively.

Since $A = A_0 \frac{L_0}{L}$, the resistance equation becomes

$$R = \rho \frac{L}{A_0} \frac{L}{L_0},$$

$$R = R_0 \left(\frac{L}{L_0} \right)^2,$$

where R_0 is the initial resistance.

c. Quadratic growth in resistance: the resistance increases quadratically with stretching due to the combined effect of increasing length and decreasing cross-sectional area:

$$R \propto \left(\frac{L}{L_0} \right)^2 = \lambda^2.$$

Problem 2. A parallel-plate stretchable capacitor is made of a soft dielectric elastomer with a relative permittivity $\varepsilon_r = 4$ and Young's modulus $Y = 400\,\text{kPa}$. The capacitor is initially unstrained and has plate area A_0, thickness t_0, and is charged to a constant voltage of $V_0 = 100\,\text{V}$.

The elastomer is then stretched uniformly in-plane to twice its original area ($A_1 = 2A_0$). Due to the nearly incompressible nature of elastomers, assume the volume of the dielectric remains constant during deformation.

a   Derive an expression for the new thickness t_1 of the dielectric after stretching, assuming volume conservation.

b   Using the parallel-plate capacitor formula:

$$C = \varepsilon_0 \varepsilon_r \frac{A}{d} \tag{2.16}$$

calculate the new capacitance C_1 relative to the original capacitance C_0.

c   Assuming the capacitor was disconnected from the power source after being charged (so charge is conserved), what is the final voltage V after stretching?

Hints:

Volume conservation: $A_0 \times t_0 = A_1 \times t_1$.

Charge conservation: $Q = C_0 \times V_0 = C_1 \times V_1$.

Solution. a   New Thickness Using Volume Conservation Let $A_0 \times t_0 = A_1 \times t_1$. Since $A_1 = 2A_0$, we have: $A_0 t_0 = 2A_0 \cdot t_1 \Rightarrow t_1 = \frac{t_0}{2}$.

b   New Capacitance Relative to Original Capacitance of a the original parallel plate capacitor is: $C_0 = \varepsilon_0 \varepsilon_r \frac{A_0}{t_0}$.

New capacitance:

$C_1 = \varepsilon_0 \varepsilon_r \frac{2A_0}{t_0/2} = \varepsilon_0 \varepsilon_r \cdot \frac{2A_0 \cdot 2}{t_0} = 4 \cdot \varepsilon_0 \varepsilon_r \frac{A_0}{t_0} = 4C_0$.

So, the new capacitance is 4 times the original:

$C_1 = 4C_0$.

c   Final Voltage After Stretching (Charge is Conserved) Conserved charge: $Q = C_0 \times V_0 = C_1 \times V_1$ Solving for final voltage V: $V_1 = \frac{C_0 V_0}{C} = \frac{C_0 V_0}{4C_0} = \frac{V_0}{4} = \frac{100}{4} = 25\,\text{V}$.

Problem 3. A dielectric elastomeric foam capacitive sensor is operated as a pressure sensor in compression. The initial sensor foam dimensions are: $1\,cm \times 1\,cm$ and $600\,\mu m$ thickness. At rest, the foam open space occupies 66 % of the total volume, whereas the elastomer the other 50 %. As the foam is compressed, the pores close, and the capacitance of the entire system changes, allowing it to be used as a pressure sensor. Determine the following:

1. The initial capacitance of the sensor if the elastomer has a dielectric constant of 3.
2. The volume fraction occupied by the elastomer within the entire foam when the sensor is compressed to $400\,\mu m$.
3. The capacitance at state 2 when total thickness is $400\,\mu m$.
4. The volume fraction occupied by the elastomer within the entire foam when the sensor is compressed to $200\,\mu m$.
5. The capacitance at state 4 when total thickness is $200\,\mu m$.

3 Fluid powered machines

Contents

Fluid powered machines are highly popular due to both the simple fabrication and ease of integration with readily available pumps and valves. This chapter focuses on the three primary methods of controlling deformation: pneumatic networks, fiber reinforced tubes, and McKibben actuators.

3.1 Learning objectives

The main concept to be presented and understood is the relationship between structure and actuation performance in systems that are inflated and deflated by a moving fluid. For all fluid systems, the structure is the primary means of directing deformation. Three primary assembly methods are described: pneumatic networks, fiber-reinforced tubes, and McKibben actuators. Trade-offs between system complexity, fabrication speed, and actuator performance are presented in the context of soft robotic applications.

3.2 Background and principles

Despite the recent widespread interest in soft robotics, the field of soft machines is fairly old and well established. Patents on various configurations of sphygmomanometers (also known as blood pressure measurement devices) are nearly a century old (e. g., US patent 2118329A was filed in 1934) and describe methods of measuring blood pressure with inflatable cuffs. Fully autonomous systems followed as soon as the rechargeable batteries were powerful enough to power wearable pumps, with patents in the 1960s. Dr. John McKibben himself was a physicist, who rose to prominence during the Manhattan project, being prominent enough to trigger the Trinity test. His invention, named the *Air Muscle* at the time, was aimed to improve the condition of his daughter, who had been diagnosed with polio and became paralyzed from the waist down.

https://doi.org/10.1515/9783111069418-003

The recent growth in interest for building soft machines comes from the widespread availability of tools and materials necessary to build them. Driven first by rapid molding of silicone elastomers to produce *pneumatic networks* (PneuNets), the effort has grown to incorporate a wide variety of *rapid prototyping techniques*, including 3D printing of molds and elastomeric structures, casting, dipping, lamination, extrusion, etc.

The simplest building block for a fluid powered soft machine can be considered the hollow chamber shown in Figure 3.1. The chamber has an elastomeric shell, which can withstand elastic deformation, and a strain-limiting layer attached to one side, shown in red. When pressure is applied, the fluid causes the chamber to expand, but the strain-limiting layer prevents the bottom layer from expanding. The result is the chamber bends around the strain-limiting layer, causing excess deformation in the top layer. A simple mathematical interpretation shows that the energy added by the fluid needs to be equal to the elastic energy stored in the elastomer. For this example, we can assume that the pressure is constant and the chamber increases in volume:

$$p \times \Delta V_{\text{chamber}} = \lambda_{\text{elastomer}} \times Y_{\text{elastomer}} \times V_{\text{elastomer}}. \tag{3.1}$$

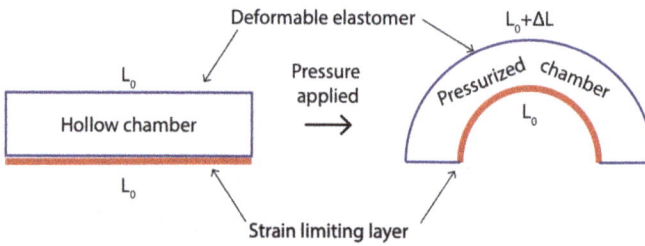

Figure 3.1: Schematic of bending deformation of a hollow chamber that has a strain limiting layer (in red) attached to one side. When pressure is applied, the chamber expands and bends around the strain limiting layer.

For simplicity and to avoid confusion, we will write any chamber components as V_c and any elastomer components with the subscript as V_e. Using the terms described in the figure, the equation becomes

$$p \times \Delta V_c = \frac{L_0 + \Delta L}{L_0} \times Y_e \times V_e = (L_0 + \Delta L)Y_e A_0 d_0, \tag{3.2}$$

where
- p is the pressure applied to the hollow chamber,
- ΔV is the change in volume for the chamber between the rest and pressurized states,
- L_0 is the initial length of the top segment of the chamber,
- ΔL is the change in length to accommodate the pressure in the top segment,
- Y_e is the area of Young's modulus in the elastomer, which forms the hollow chamber,
- A_0 is the area of the top segment in the initial state, and

– d_0 is the thickness of the top segment in the initial state.

The parametric space for understanding the performance of fluid powered machines is contained within equation (3.2). Pressure and volume changes are linearly proportional to the amount of elastic energy stored in the elastomeric material. The geometric parameters of the deformable chamber each impact the response in the material. Often, the radius of curvature of the structure is important to the final application. From bending beam theory we can find a simple approximation for relatively thin chamber bending:

$$\frac{L_0 + \Delta L}{L_0} = \kappa \frac{d_0}{2},$$

(3.3)

where κ is the curvature of the beam. The following sections discuss specific configurations of elastomeric chambers and strain-limiting layer designed to produce bending, contraction, expansion, or twisting deformation.

3.3 PneuNets: pneumatic networks

PneuNets, or pneumatic network actuators, are structures comprised of multiple chambers, each capable of expansion against each other. The chambers are designed with thinner walls where more expansion is desired. The most common PneuNet actuator is a bender that shows the ability to change from a flat state at rest to a curved shape, up to 360 degrees along the strain limiting layer to wrap around a specific object target.

Figure 3.2 shows a schematic of a PneuNet that contains six hollow chambers, connected by a common fluid line, with an inlet. The walls of the six hollow chambers are significantly thinner than the elastomeric caps, and a strain-limiting layer, shown in red, is adhered onto the elastomer around the fluid channel, opposite to the hollow chambers. When a pressure is applied through the fluid inlet, each chamber begins to expand, and the thinner walls cause preferential expansion from each chamber against its neigh-

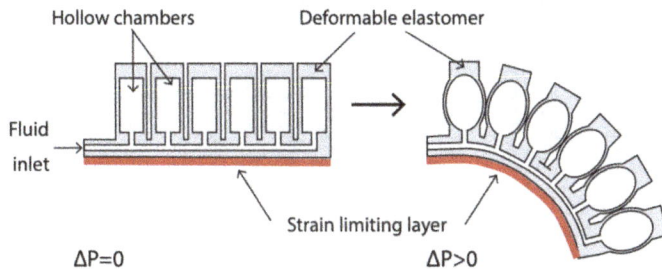

Figure 3.2: Diagram showing the deformation of a PneuNet actuator with six expandable chambers where the side walls are thinner than the caps, causing preferential deformation where the chambers push against each other. Overall, applied pressure causes the actuator to bend downward relative to the initial state.

bors. The combined effect of the chamber expansion and presence of a strain limiting layer causes the PneuNet actuator to bend downward, curving the overall structure. The natural bending motion makes PneuNet actuators particularly useful for gripper-type applications and crawling locomotion.

The main challenge, and innovation, in the development of PneuNets was fabrication of complex 3D structures from soft materials. The original demonstrations and most established fabrication approaches rely on casting two segments of the actuator separately, then bonding the elastomeric structures to produce the cavities and fluid channel. The cast part, which includes the hollow chambers, has overhangs and complex features, but molding from siloxane elastomers manages to retain features, and the soft material can be deformed around the mold to detach easily. The strain limiting layer is cast together with the bottom half of the mold, as shown in Figure 3.3, and can be made from paper. As discussed in Chapter 1, the paper's microstructure allows the

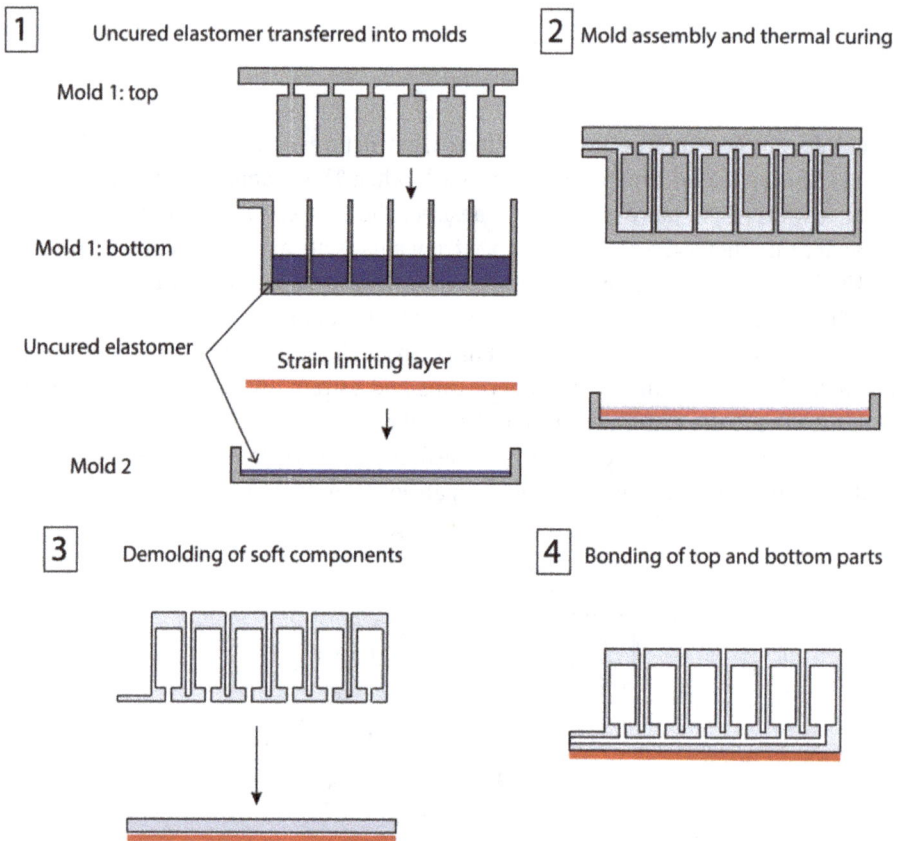

Figure 3.3: Steps in the assembly of a PneuNet actuator: 1. Uncured elastomer is poured into molds; 2. Molds are assembled, and elastomer is cured thermally; 3. The two elastomeric parts are demolded; 4. The PneuNet is assembled by bonding the two cast parts.

silicone elastomer to permeate through while still uncured and in liquid form. During the curing process, the paper is embedded into the silicone layer and is mechanically interlocked. Lastly, the two cast parts are bonded together, usually with a silicone adhesive, or uncured silicone of the same chemical nature as the main body. Often, fully cured silicone parts show limited adhesion to uncured material, making the seam between the two parts one of the weakest points, and potential failure location. Multiple techniques have been developed to mitigate this known weakness.

3.4 Fiber-reinforced actuators

Fiber-reinforced actuators are hollow enclosures made of elastomer surfaces reinforced with a network of fibers. Exact mathematical models have been developed to capture the behavior of these actuators in response to applied pressure as a function of the geometric parameters of the fibers. The main finding was that the kinematics of the actuators are completely controlled by the fiber orientations.

Considering the example actuator in Figure 3.4, several assumptions can be made to better understand the behavior of the fiber-reinforced actuator. First, the fibers that provide the reinforcement are assumed to be inextensible. Mathematically, this is written as

$$\lambda_1^2 \cos(\alpha)^2 + \lambda_2^2 \sin(\beta)^2 \left(\frac{\theta^*}{\theta} \right)^2 = 1, \tag{3.4}$$

where
- λ_1 is the axial strain,
- λ_2 is the radial strain,
- α is the angle of the first set of fibers relative to the cylinder axis,

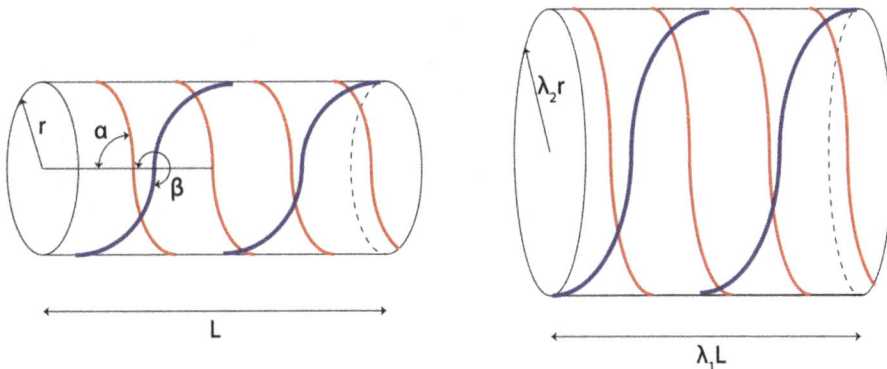

Figure 3.4: Schematic of a fiber-reinforced cylinder of radius r and length L with two sets of fibers, aligned at angles α and β. On the right side, the cylinder is deformed to a larger radius $\lambda_2 r$ and smaller length $\lambda_1 L$.

- β is the angle of the second set of fibers relative to the cylinder axis,
- θ is the number of rotations of the fibers in radians before deformation, and
- θ^* is the number of rotations of the fibers in radians after deformation.

Similarly, the volume change between states depends on the strain applied, which is expressed as

$$V_* = \lambda_2^2 \lambda_1 \pi r^2 l. \tag{3.5}$$

The total number of rotations in the fiber depends on the geometry of the cylinder and the angle of the fiber

$$\theta = \frac{l}{r} \tan(\alpha). \tag{3.6}$$

The model, so far, makes no assumptions about how the cylinder is deformed, which makes the method generalizable to other types of actuation, as long as the result is an axial and radial strain. The model can be used to predict the deformation types in an actuated cylinder for specific strains to produce three distinct deformation modes: rotation between actuated and unactuated, extension (or contraction) relative to the amount of rotation, and bending of the entire structure.

The angle δ gives the rotation in the actuated state relative to the unactuated state and is found with

$$\theta = \frac{l}{r} \frac{\frac{\beta}{|\beta|} \sin(\alpha)\sqrt{1 - \lambda_1 \cos(\beta)^2} - \frac{\alpha}{|\alpha|} \sin(\beta)\sqrt{1 - \lambda_1 \cos(\alpha)^2}}{\frac{\alpha}{|\alpha|} \cos(\beta)\sqrt{1 - \lambda_1 \cos(\alpha)^2} - \frac{\beta}{|\beta|} \cos(\alpha)\sqrt{1 - \lambda_1 \cos(\beta)^2}}. \tag{3.7}$$

The result of the rotation equation can be used to calculate the pitch (extension per rotation) of the fiber-reinforced actuator during the actuation, to find the closed-form solution

$$p = r \frac{\sin(\alpha) \sin(\beta) \sin(\alpha - \beta)}{\sin^2(\alpha) - \sin^2(\beta)}. \tag{3.8}$$

The bending deformation is more complex and not easily captured by the closed-form analytical model developed by Kota et al. (Bishop-Moser and Kota, "Design and modeling of generalized fiber-reinforced pneumatic soft actuators"). However, the force due to an applied pressure can be written as

$$F = P\pi r^2 (1 - 2\cot^2(\chi)), \tag{3.9}$$

where χ is on of the angles between α and β, which is further from the axial direction.

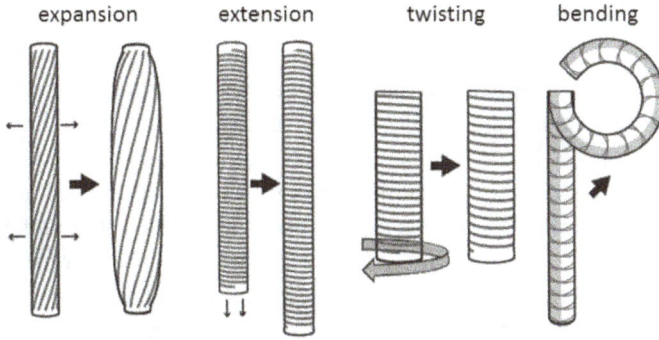

Figure 3.5: The four modes of deformation that emerge from fiber-reinforced cylinders: axial expansion, linear expansion, twisting, and bending. Reproduced from the Soft Robotics Toolkit.

The broad range of movements that emerge from this simple construction is captured in Figure 3.5. The specific angles for each deformation mode are discussed in detail in the research literature, capturing unique conditions, such as locking angles where no movement occurs and critical angles at which transitions happen.

Similarly to PneuNets, a longstanding challenge in fiber-reinforced actuators is the fabrication of the devices. The fibers need to be integrated into the structure but cannot bond to the elastomer in a way that restricts its deformation. A sequence of steps is shown in Figure 3.6 as an example to illustrate the complexity of the process needed to produce a single half round bending actuator.

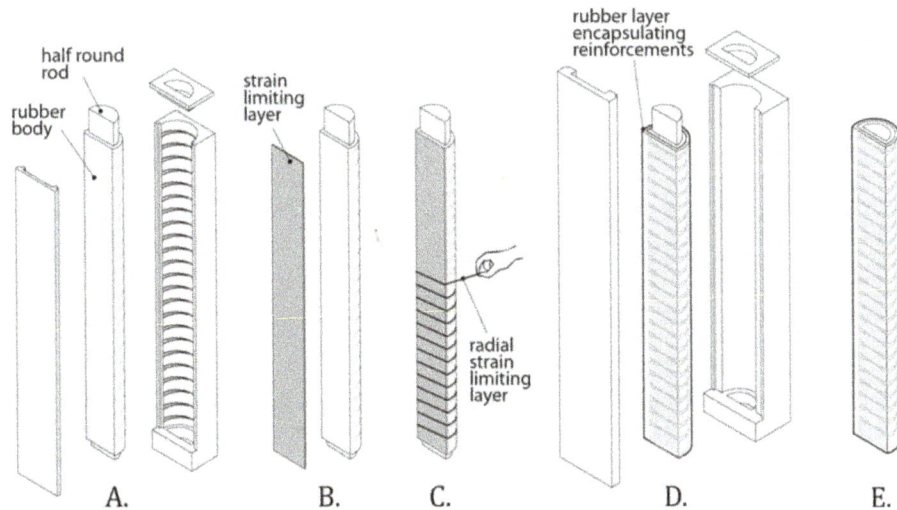

Figure 3.6: Sequence of steps needed to produce a fiber-reinforced actuator: A. Casting the elastomeric body; B. Adding a strain limiting layer and fiber reinforcements; C. Encapsulating the strain limiting elements. Reproduced from the Soft Robotics Toolkit.

3.5 McKibben actuators

McKibben actuators are a specific subset of fiberreinforced actuators, which produce reliable expansion or contraction movement depending on the angle of the fibers. One of the earliest configurations, also known as pneumatic artificial muscles (PAMs), consists of a hollow elastomer tube reinforced with two families of fibers, with one wound in a clockwise helix (CW) and the other wound in a counterclockwise (CCW) helix of the same pitch. Figure 3.4 shows how radial expansion of the chamber causes linear axial contraction, allowing operation as an artificial muscle. The interest in and development of McKibben-type actuators is primarily due to the ease of deriving geometric relationships that govern the deformation behavior of the actuators.

In a simple mathematical model, the performance of McKibben actuators is controlled by one input: the applied pressure P and two main design parameters, the angle of the reinforcements, or braids ($\alpha = \beta$), and the cross-sectional area of the actuator (A), and the Young's modulus of the elastomeric material (Y). The output force in a McKibben actuator can be expressed as

$$F = \pi r_1^2 P \frac{1}{\sin^2 \alpha}\left(3\left(1 - \frac{L_1 - L_2}{L_1}\right)^2 \cos^2 \alpha - 1\right). \tag{3.10}$$

Multiple improved models have been developed to capture the impact of the reinforcement fibers and their specific properties, but this simple model captures the proportionality relations between the key parameters. The most attractive feature of the McKibben actuators is the contractile response to pressure for a broad range of fiber angles, as exemplified in Figure 3.7. The contraction mimics the behavior of natural muscles and earned the McKibben actuators the secondary name of pneumatic artificial muscles.

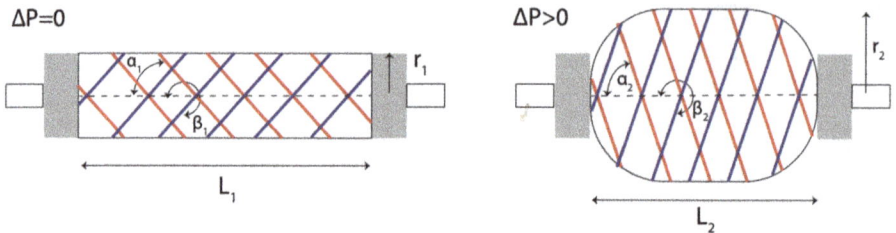

Figure 3.7: Operation of a McKibben actuator, where the fibers are aligned at angles equal in value but opposite sign: when pressurized, the chamber inflates, causing the length of the actuator to decrease.

Across multiple disciplines, significant progress has been made in the field of fluid-driven machines by pushing the boundaries of design and manufacturing. The main limitation of these systems comes from the *integration* aspect: all are powered by fluids that require pumps and valves to cause deformation used in locomotion and manipulation. Most readily available pumps are rigid and heavy, negating the advantages of soft machines at the system level. The following chapters introduce other types of machines, powered by thermal, electrical, and magnetic energy, which alleviate some of the challenges in powering soft machines. In hybrid approaches, fluids can be moved by soft pumps powered by electrical, thermal, or magnetic energy to power or control soft machines and make the entire system softer and more compliant.

Additionally, the structure and operation mode of fluid-driven soft machines do not readily lend themselves to *sensing* abilities. For machines to operate autonomously, they require some knowledge of their own state and environment. Sensing strain, temperature, pressure, proximity, and the presence of chemical species are all abilities that natural systems have and that can be added to rigid machines. Adding sensors to fluid-driven machines has been done extensively, but it comes with an additional integration cost: most commonly, the sensors need electrical power and provide electrical signals that need to be interpreted by separate electronic systems. Again, advances in other types of soft machines have given rise to a rich space of hybrid approaches, increasing the range of robotic capabilities.

3.6 Alternative approaches: combustion power and eversion robots

There are two modification of fluid-driven soft machines that are worth mentioning for their unique aspects: combustion powered systems and eversion robots. Given the challenge of pumping fluids to power the soft machines, some researchers envisioned the use of fuels that convert into gases to power the robots.

In one example, butane and oxygen are fed into the reaction chamber of a purpose-made robot and detonated with a spark. The resulting reaction, oxidation of butane, produces carbon dioxide and water vapor, which expand rapidly to cause impulsive movement in the robot. As shown in Figure 3.8, this explosion can be used to cause jumping locomotion in a centimeter-scale robot:

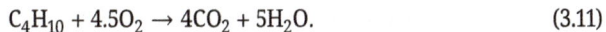

$$C_4H_{10} + 4.5O_2 \rightarrow 4CO_2 + 5H_2O. \tag{3.11}$$

Although combustion has its benefits, primarily in enabling impulsive movements, the amount of energy produced can be damaging to the robot. An alternative way of producing a working gas is decomposition of a liquid, such as hydrogen peroxide. The

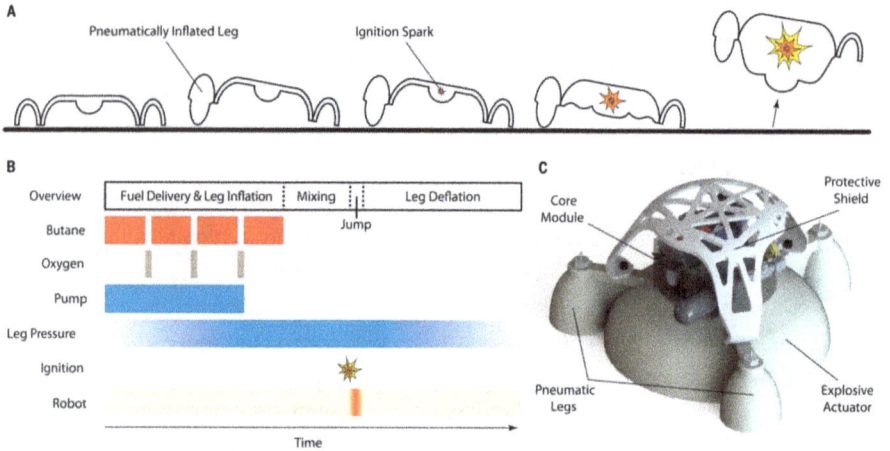

Figure 3.8: From Bartlett et al. "A 3D-printed, functionally graded soft robot powered by combustion": Robot design and principle of operation. (A) To initiate a jump, the robot inflates a subset of its legs to tilt the body in the intended jump direction. (B) The ignition sequence consists of fuel delivery, mixing, and sparking. Butane and oxygen are alternately delivered to the combustion chamber (to promote mixing). (C) Computer-aided design model of the entire robot, consisting of the main explosive actuator surrounded by three pneumatic legs.

decomposition is explained in the equation below and will be discussed more in the chapter on *Advanced Topics*. From a machine design perspective, this system is useful because it runs on a chemical fuel (e. g., H_2O_2), produces a gas (O_2), and the decomposition rate depends on the type of catalyst used, giving the operator some control over the entire process.

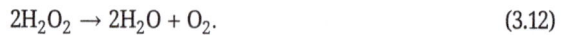

$$2H_2O_2 \rightarrow 2H_2O + O_2. \tag{3.12}$$

Separately, eversion-type robots have a unique method of navigating their environments through growth, as shown in Figure 3.9. Often, the driving force for movement is an applied pressure, which links these eversion robots conceptually to the broader field of fluid-powered machines. The eversion robots, first described in 2017 "A soft robot that navigates its environment through growth", are a class of soft pneumatic robots capable of growing substantially in length from the tip while actively controlling direction. The eversion behavior is controlled by two inputs: directional growth via pressurization of an inverted thin-walled vessel allows for rapid and substantial lengthening of the tip of the robot body and steering via controlled asymmetric lengthening of the tip.

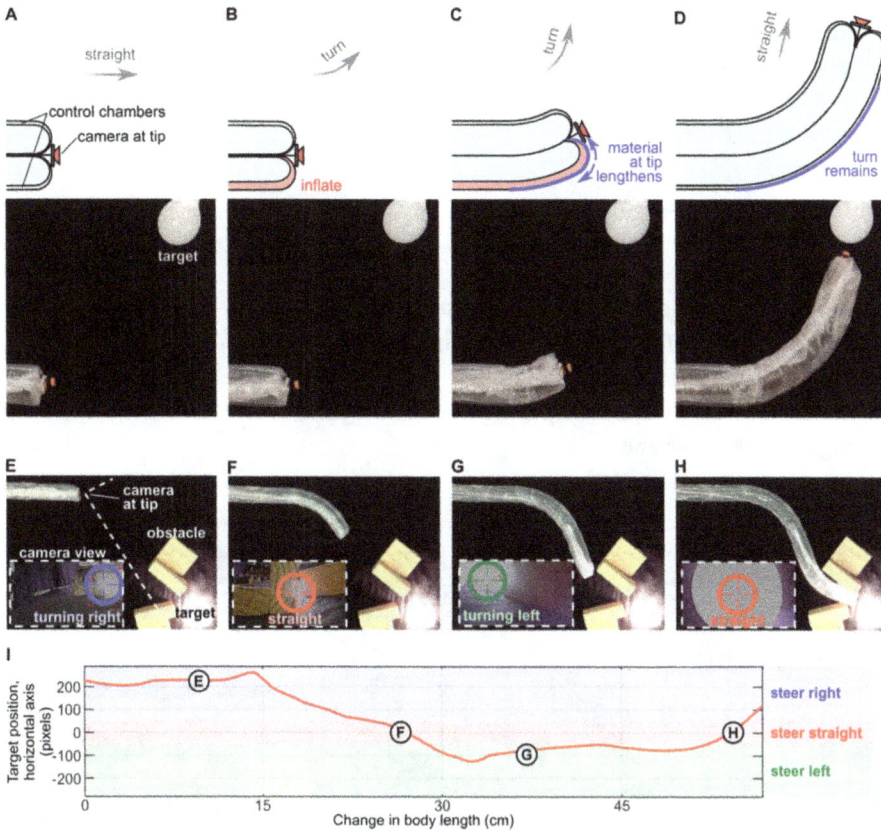

Figure 3.9: From Hawkes et al: Principle of asymmetric lengthening of tip enables active steering. (A) Implementation in a soft robot uses small pneumatic control chambers and a camera mounted on the tip for visual feedback of the environment. (B) To queue an upward turn, the lower control chamber is inflated. (C) As the body grows in length, material on the inflated side lengthens as it everts, resulting in an upward turn. (D) Once the chamber is deflated, the body again lengthens along a straight path, and the curved section remains. (E) A soft robot can navigate toward light using a tip-mounted camera.

3.7 Example laboratory session A: PneuNets

In this laboratory session, students will test fluid-powered actuators materials, PneuNets, for both deformations. The goal of the lab is to give them an understanding of how energy is converted from a moving fluid into components capable of locomotion or manipulation.

The fabrication station will include all of the materials shown in Figure 3.10:
- Eco-Flex 00-35 FAST Part A,
- Eco-Flex 00-35 FAST Part B,
- mold release spray,
- syringe,
- tweezers,
- scissors,
- scale,
- flat mold for strain limiting layer with paper,.
- two-part mold to fabricate the pneumatic chambers,
- Luer lock connector,
- syringe needles to make an opening for the Luer lock,
- digital angle ruler, and
- lab stands with gripper.

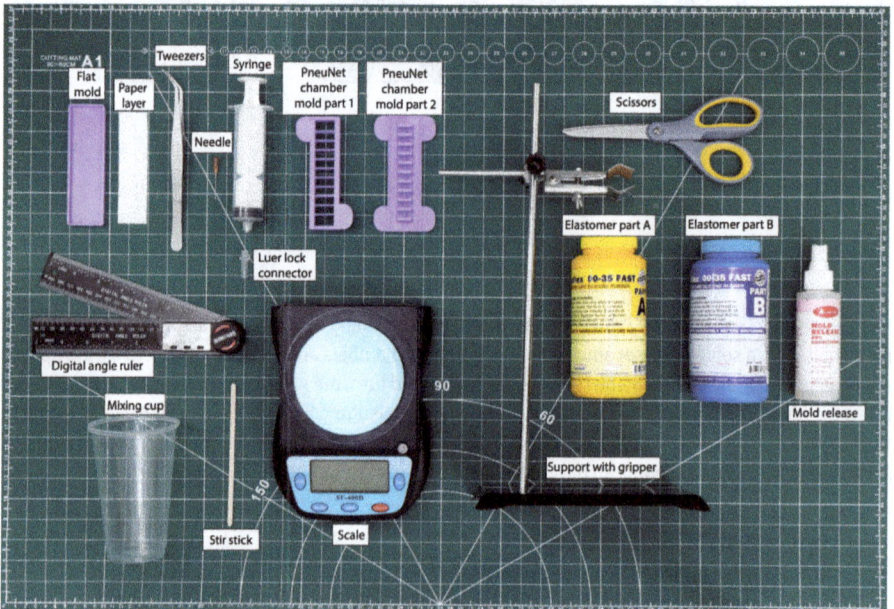

Figure 3.10: Preparation for laboratory example 3A: all of the materials required to make PneuNet actuators.

Figure 3.11 outlines the first steps in building a PneuNet actuator, which involve weighing and mixing a silicone elastomer (**Steps 1–3**), similar to the previous two labo-

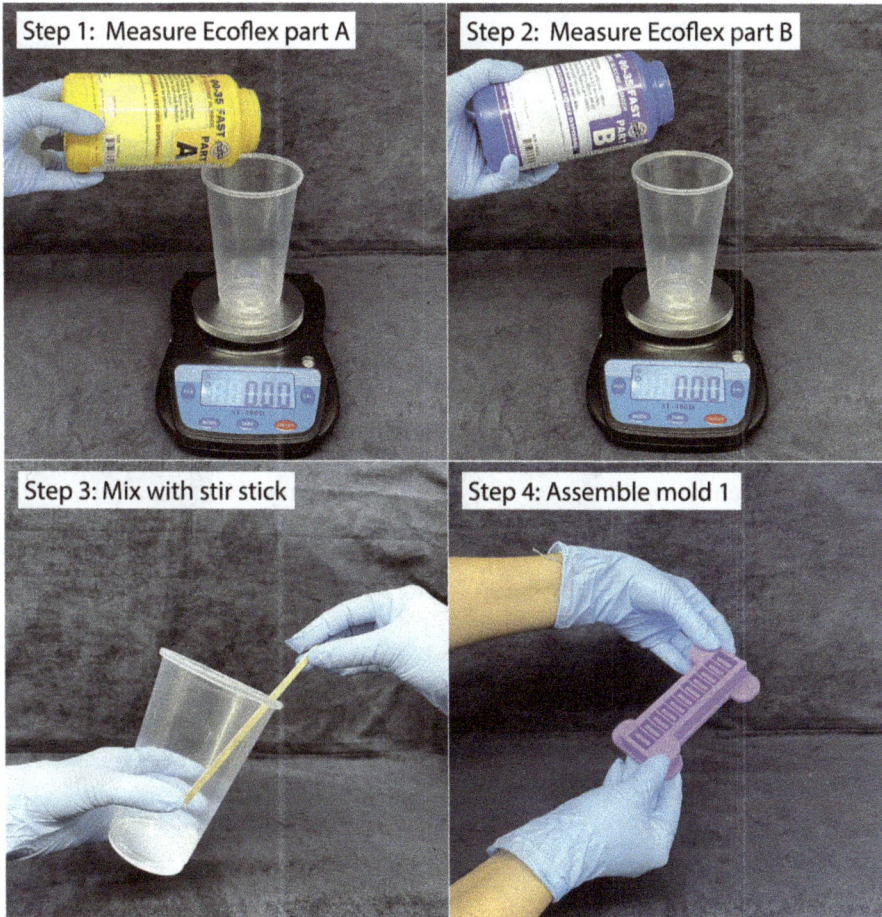

Figure 3.11: Steps 1–4 in a PneuNet assembly.

ratory sessions. The two-part mold, which produces the PneuNet chambers, needs to be assembled correctly to make evenly spaced pneumatic network chambers (**Step 4**).

Mold release solution should be added to the chambered mold to make removing the cured part easier (**Step 5**). Figure 3.12 shows how the elastomer is added to the chambered mold (**Step 6**), then allowed to cure and removed from the mold (**Step 7**). Separately, the strain limiting layer, in this case filter paper, is pressed into the second flat mold (**Step 8**).

The following steps are shown in Figure 3.13 starting with more elastomer is added to the flat mold (**Step 9**), then allowed to partially cure, similarly to how a channel is made in the Stretchable Electrode laboratory. The completed chambered part is added

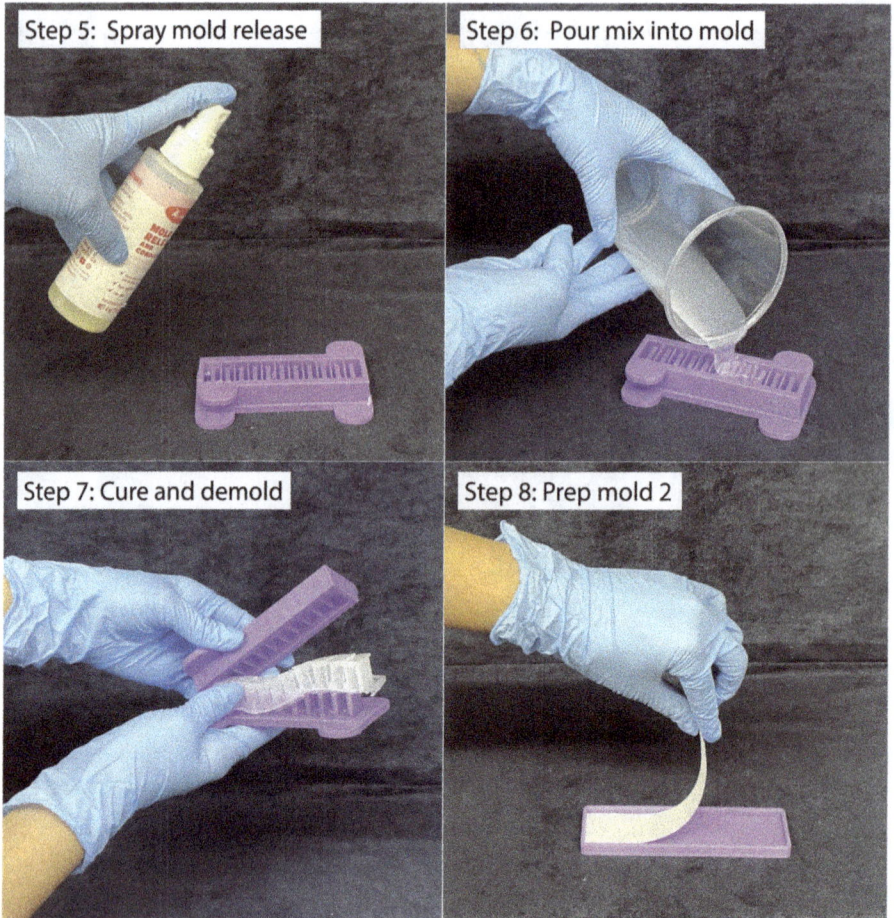

Figure 3.12: Steps 5–8 in a PneuNet assembly.

to the flat mold with partly cured elastomer to ensure good bonding between the two parts (**Step 10**). Once the elastomer is fully cured, the entire PneuNet is removed from the mold (**Step 11**). The process of making reliable connections to the inner chamber of the PneuNet involves piercing the elastomeric shell with the given needle (**Step 12**).

As shown in Figure 3.14, the final steps in the PneuNet testing include inserting a barbed Luer lock connector through the needle piercing hole (**Step 13**). The connector can then be secured in place with more uncured elastomer. The syringe is then connected to the Luer lock part (**Step 14**) and pressurizes the PneuNet to produce curving deformation (**Steps 15–16**).

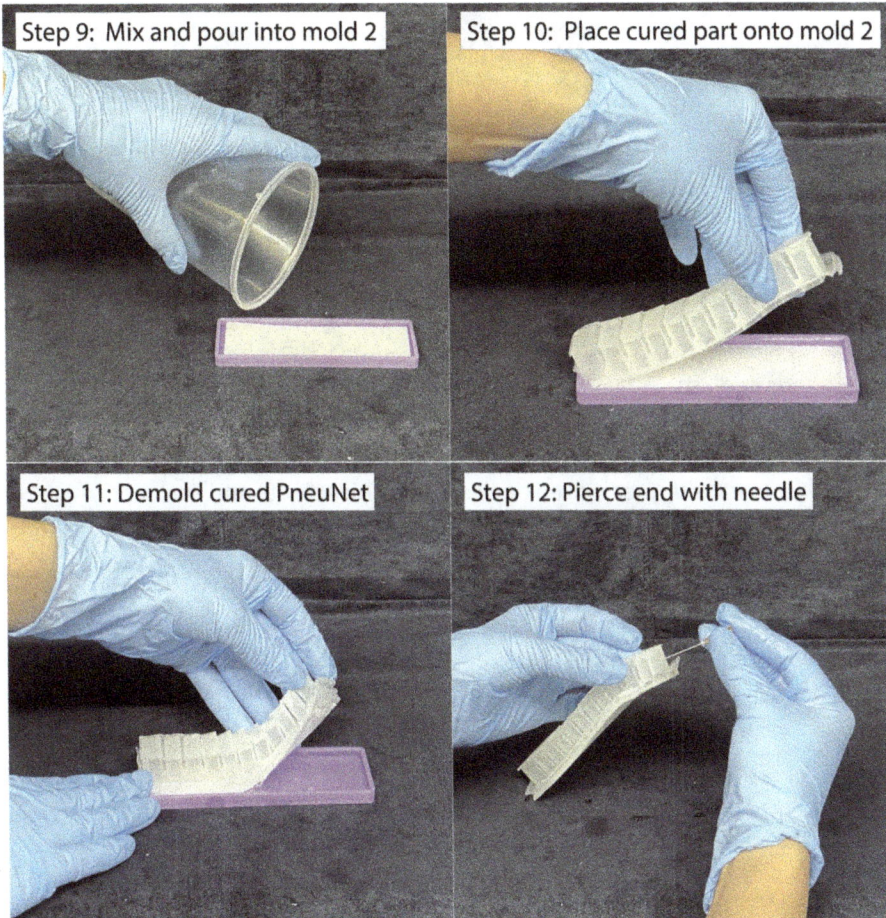

| Step 9: Mix and pour into mold 2 | Step 10: Place cured part onto mold 2 |
| Step 11: Demold cured PneuNet | Step 12: Pierce end with needle |

Figure 3.13: Steps 9–12 in a PneuNet assembly.

With 3D printing ability, different size molds can be made to produce PneuNets of different starting lengths. The students can aim to test one of each and share devices with other teams once they complete a set of tests, evaluating radius of curvature as a function of elastomer properties, system size, and applied pressure as shown below.

Elastomer	Length (cm)	Pressure (kPa)	Radius of curvature (cm)

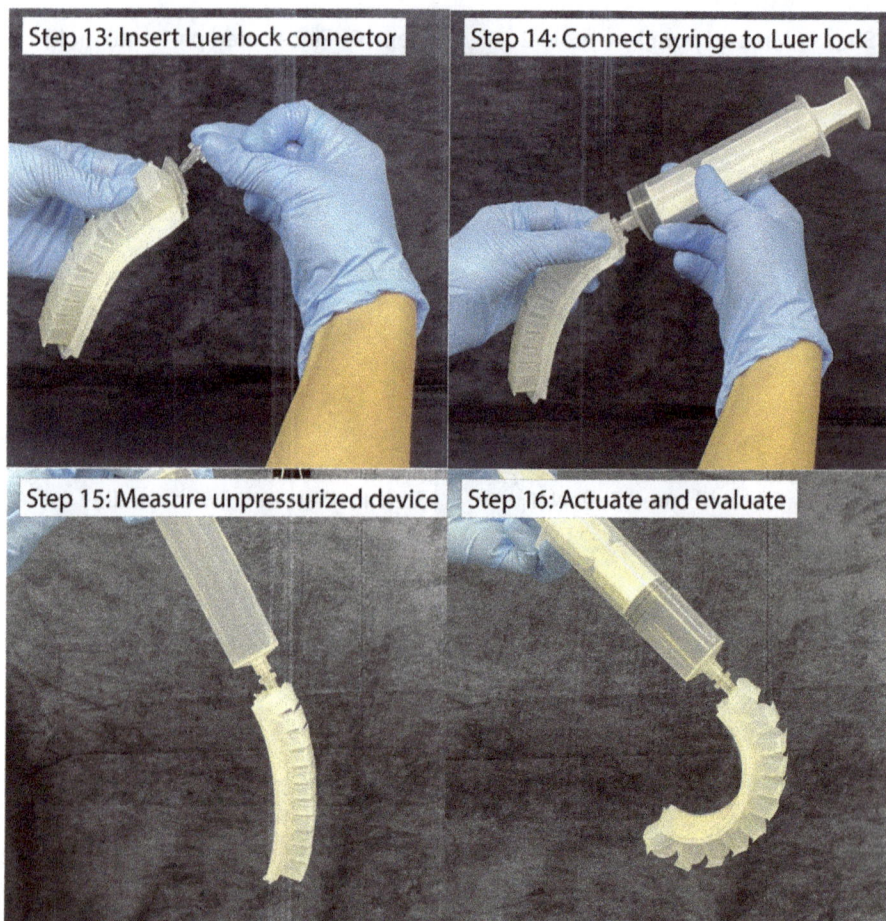

Figure 3.14: Steps 13–16 in a PneuNet assembly.

3.8 Example laboratory session B: McKibben actuators

In this second laboratory session, students will test fluid-powered actuator materials: McKibbens for force and displacement. The goal of the lab is to give them an understanding of the envelope of force vs. displacement, as shown in the last figure in the chapter.

The McKibben fabrication station will include all of the materials shown in Figure 3.15:

- syringe,
- rigid tubing,
- zipties,

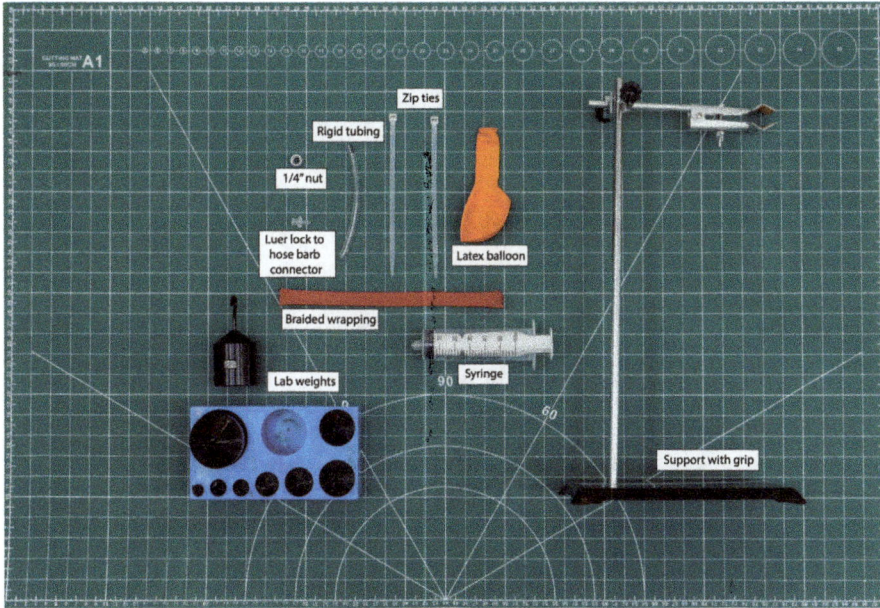

Figure 3.15: Preparation for laboratory example 3B: all of the materials required to make McKibben actuators.

- tweezers,
- metal nut to secure braid,
- scissors,
- balloon,
- braided wrapping,
- Luer lock connector,
- ruler,
- lab stands with gripper, and
- lab weights.

For this section, there are balloons of different lengths from which McKibben actuators can be made. Aim to test one of them and share devices with other teams, once you have completed a set of tests.

To build a McKibben, the students will pass braided wrapping through metal nut (**Step 1**). Next, the braid is secured with a zip tie, and the excess material is cut off (**Step 2**). If available as an alternative, the braid can be melted with a heat gun to avoid the need for a zip tie. Then the students can place a balloon inside braid, with the opening opposite to the ziptied or melted closed side (**Step 3**). The last step in Figure 3.16 shows how to insert the rigid tubing into the balloon (**Step 4**).

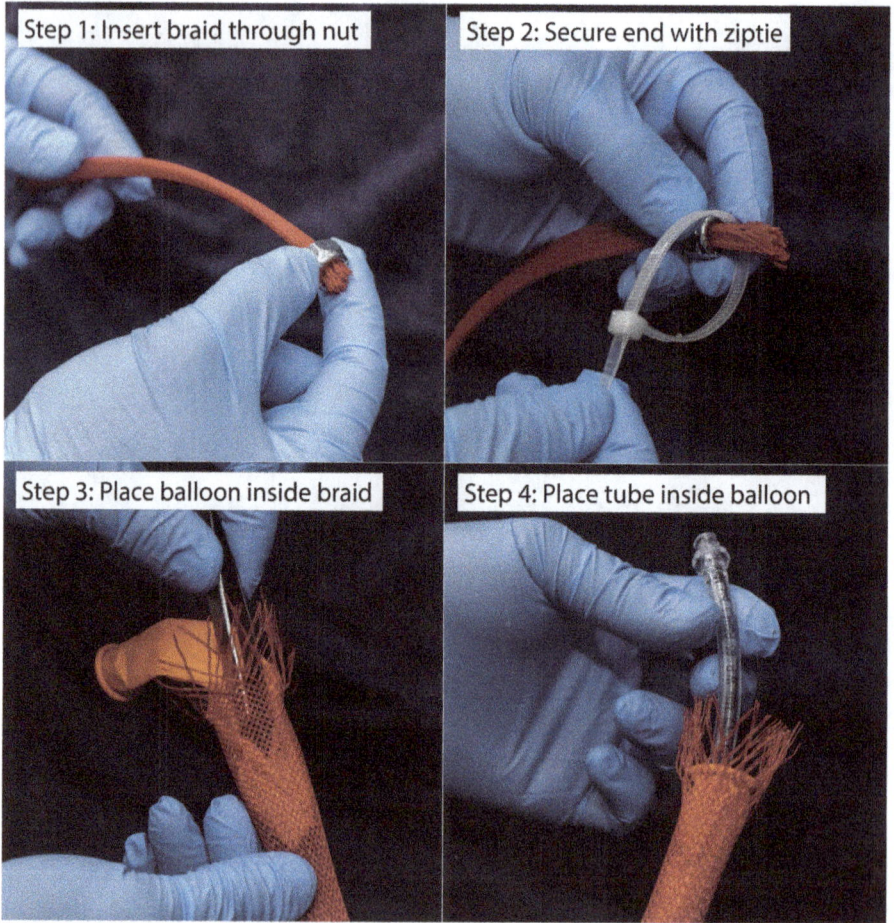

Figure 3.16: Steps 1–4 in a McKibben assembly.

Figure 3.17 shows how the McKibbens are completed and tested. The braid and balloon are secured with a second ziptie, and then the excess material is cut off (Step 5). To allow for pressurization via syringe, the students then connect the barb hose ending to the rigid tubing and connect the syringe to the other end of Luer lock connector (**Step 6**). To measure deformation, the next step is to secure the rigid tubing on the lab stand and add weights to the metal nut (**Step 7**). The main action item for the students is to pressurize the syringe and record how much the weight is lifted as a difference in centimeters. For the laboratory report, the students can record results in the table below, aiming to map out a figure similar to Figure 3.18.

Balloon length (cm)	Pressure (kPa)	Weight (g)	Height (cm)

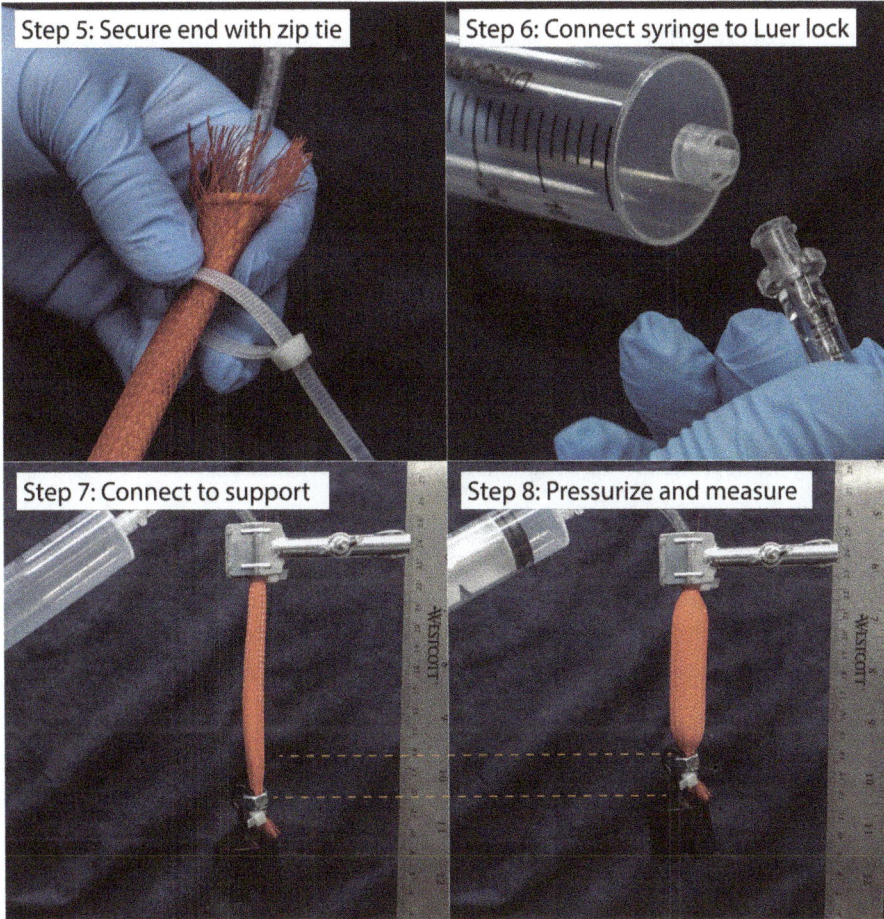

Figure 3.17: Completing the assembly and testing a McKibben.

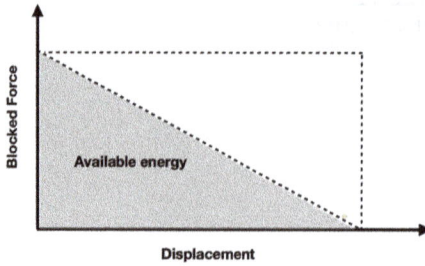

Figure 3.18: Example of a force vs. displacement characterization of an actuator, to be completed with laboratory data for specific McKibben examples.

3.9 Example problems

Problem 1. A McKibben actuator is a type of pneumatic artificial muscle that contracts when pressurized. The force it generates depends on the internal air pressure and the geometry of the actuator, particularly the diameter of the tubing.

Suppose you have two McKibben actuators: Actuator A has an initial (unpressurized) diameter of d_A = 20 mm; actuator B has an initial diameter of d_B = 30 mm.

Both actuators have the same braided fiber angle (θ = 25°) and length (L = 200 mm) and are tested at pressures P = 100 kPa and P = 200 kPa.

Using the approximate force equation for McKibben actuators

$$F = PA(3 \cos^2 \theta - 1),$$

where $A = \frac{\pi d^2}{4}$ is the cross-sectional area of the actuator, determine: 1. The force exerted by each actuator at 100 kPa and 200 kPa; 2. How the force changes with pressure and tubing diameter.

Solution. 1. Calculate the cross-sectional area for each actuator:

$$A_A = \frac{\pi(0.02)^2}{4} = 3.14 \times 10^{-4} \, m^2,$$

$$A_B = \frac{\pi(0.03)^2}{4} = 7.07 \times 10^{-4} \, m^2.$$

2. Calculate the force for each actuator at 100 kPa:

$$F_A = (100 \times 10^3) \times (3.14 \times 10^{-4}) \times (3 \cos^2 25° - 1)$$

$$\approx 100000 \times 3.14 \times 10^{-4} \times 1.56$$

$$\approx 49 \, N,$$

$$F_B = (100 \times 10^3) \times (7.07 \times 10^{-4}) \times 1.56$$

$$\approx 110 \, N.$$

3. Calculate the force for each actuator at 200 kPa. Since force scales linearly with pressure,

$$F_A \text{ at } 200 \text{ kPa} = 2 \times 49 = 98 \text{ N},$$
$$F_B \text{ at } 200 \text{ kPa} = 2 \times 110 = 220 \text{ N}.$$

4. Observations: Increasing pressure doubles the force output. A larger tubing diameter results in a greater cross-sectional area, which significantly increases the force output. Actuator B (with a 50 % larger diameter) produces more than twice the force of actuator A.

Problem 2. A cylindrical soft actuator consists of an elastomeric chamber reinforced with inextensible fibers helically wrapped at a constant angle θ (measured with respect to the longitudinal axis). When pressurized internally with fluid at pressure P, the actuator expands and deforms. The assumptions are:

- The actuator has an initial length L_0 and radius R_0.
- The elastomeric material is incompressible and isotropic.
- The fibers do not stretch and fully constrain radial and axial expansion based on their wrapping angle.
- Axial deformation dominates over radial deformation.
- The internal volume remains cylindrical during actuation.

Determine the following:
1. Derive an expression relating the axial strain λ_z of the actuator to the fiber angle θ.
2. For fiber angles $\theta = 30°, 45°$, and $60°$, estimte the expected axial strain as a function of applied pressure.
3. Explain how changing the fiber angle affects whether the actuator primarily elongates, contracts, or twists when pressurized.

Solution. Part 1: Derivation of the relationship between strain and angle: The key idea is that the inextensible fibers constrain the geometry of the actuator during pressurization. For a helical fiber wrap, the relationship between the actuator's geometry and the fiber angle θ is preserved. For the solution, L_1 is the length after deformation, R_1 is the new radius, and n is the number of fiber turns over the initial length L_0.

Since fibers are inextensible, their length remains constant:

$$l_f = \sqrt{(2\pi nR_1)^2 + L_1^2} = \text{constant}$$

We can define the initial angle of the fibers:

$$\tan\theta = \frac{2\pi nR_0}{L_0} \Rightarrow n = \frac{L_0 \tan\theta}{2\pi R_0}$$

Now using the constancy of fiber length and assuming incompressibility (volume conserved, so $R_1^2 L_1 = R_0^2 L_0$), we can derive the relationship between the deformed length L and initial conditions:

$$l_f^2 = (2\pi n R_1)^2 + L_1^2$$

Substitute n and eliminate R_1 using incompressibility:

$$R_1 = R_0 \sqrt{\frac{L_0}{L_1}} \Rightarrow l_f^2 = \left(2\pi \cdot \frac{L_0 \tan \theta}{2\pi R_0} \cdot R_0 \sqrt{\frac{L_0}{L_1}} \right)^2 + L_1^2$$

$$l_f^2 = \left(L_0 \tan \theta \sqrt{\frac{L_0}{L}} \right)^2 + L_1^2 = L_0^2 \tan^2 \theta \cdot \frac{L_0}{L_1} + L_1^2$$

This equation can be solved numerically for L_1, given θ, L_0, and R_0. The axial strain is:

$$\lambda_z = \frac{L_1 - L_0}{L_0}$$

Part 2: Qualitative behaviors for different angles:
- $\theta = 30°$: More axial alignment; the actuator elongates significantly as fibers restrict radial growth.
- $\theta = 45°$: Balanced axial and radial restriction; moderate elongation.
- $\theta = 60°$: More circumferential alignment; axial shortening or radial bulging may occur.

Part 3: Effect of fiber angle on actuation mode
- Small angles ($\theta < 54.7°$): Fibers favor axial elongation.
- $\theta = 54.7°$: The so-called "neutral angle" where elongation and radial expansion balance.
- Large angles ($\theta > 54.7°$): Fibers favor radial expansion, resulting in axial contraction or ballooning.

If asymmetric wrapping or multiple layers are used, twisting can also emerge.

Problem 3. PneuNets can be used for adhesion, as shown in Figure 3.19, where a PneuNet system is modified to change the pressure locally at the contact of a cavity with a surface. Modifying the cavity's volume lowers the pressure and produces an adhesive force, which you have to calculate. Calculate the following:
1. What is the volume of the cavity at the start? Assume it is a spherical dome with diameter 2 cm and height 0.25 cm.
2. As the pneumatic structure is inflated above, the dome deforms and reaches a hemispherical shape with the diameter staying constant and the height increasing to 1 cm. What is the area strain at the surface of the dome?

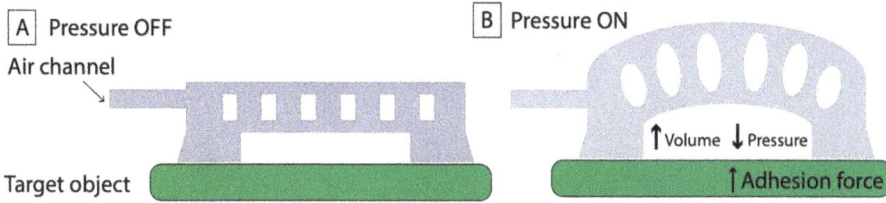

Figure 3.19: Octopus-inspired suction mechanism using a PneuNet approach.

3. What is the final volume of the dome and the local pressure, assuming that Bernoulli's principle holds?
4. An alternative way to determine the adhesion force is to calculate the change in pressure and use it and the area where the pressure is applied to find adhesive force. Use the equation

$$\Delta P = \frac{P_0 \Delta V}{V_0 + \Delta V},$$

where P_0 is the initial pressure, and V_0 is the initial volume. Assuming that the contact area does not change, find the adhesive force as the product of ΔP and area.

Bibliography

Joshua Bishop-Moser and Sridhar Kota. "Design and modeling of generalized fiber-reinforced pneumatic soft actuators". In: *IEEE Transactions on Robotics* 31.3 (2015), pp. 536–545.

Nicholas W Bartlett et al. "A 3D-printed, functionally graded soft robot powered by combustion". In: *Science* 349.6244 (2015), pp. 161–165.

Elliot W Hawkes et al. "A soft robot that navigates its environment through growth". In: *Science Robotics* 2.8 (2017), pp. eaan3028.

4 Dielectric elastomer transducers

Contents

Dielectric elastomer transducers operate as soft sensors, actuators, and energy harvesting devices. Their fundamental operation mode is electro-mechanical transduction, which can be done in either forward or reverse directions. Though conceptually simple and relatively straightforward to model, these devices are challenging to build and operate under realistic conditions.

4.1 Learning objectives

The main concepts are behavior of stretchable capacitors and energy conversion from electrical to mechanical and vice versa. An additional goal for the chapter is to link material properties to device performance, for example, evaluating the viscoelasticity of the elastomer and sheet resistance of the electrodes to understand which is the rate-limiting step for fast actuation: charging the capacitor or responding to the developing Maxwell stress.

4.2 Background and principles

A **dielectric elastomer transducer** is a capacitor in which the central dielectric is a stretchable elastomer. By pairing the elastomer dielectric with stretchable electrodes a fully compliant capacitor can be produced. When dielectric elastomer actuators (DEAs) are used as actuators, they convert input electrical energy into mechanical work. Figure 4.1 shows a schematic of the typical deformation in a parallel plate soft capacitor when an electric field is applied across the compliant electrodes.

A *stretchable electrode* is a material that can conduct electricity and undergo uniaxial or biaxial stress while remaining conductive. Stretchable electrodes were identified in the first major publication on DEAs as the key limitation of the technology and have remained a major challenge for the field.

https://doi.org/10.1515/9783111069418-004

Voltage off   **Voltage on**

Electrode

Elastomer

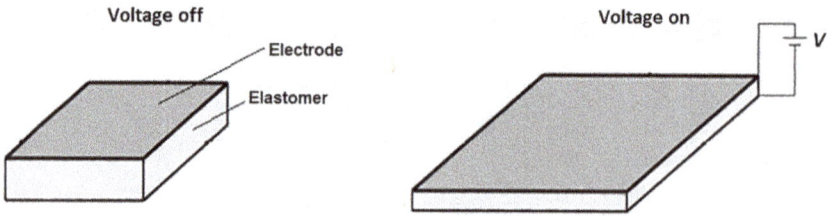

Figure 4.1: Example of a dielectric elastomer actuator, responding to applied field. Reproduced from the Soft Robotics Toolkit.

Prestretch: historically, most DEA demonstrations have been performed with elastomers that are prestretched, either uniaxially or biaxially, and attached to a rigid frame to retain the strained state. Prestretch is required for most elastomers to reach a plateau in the stress vs. strain curve, a state where small changes in the applied electric field cause large deformation.

Electromechanical instability relates to a self-reinforcing mode in which DEAs can get damaged. When an elastomer membrane is deformed by an applied electrical field, it becomes thinner, which in turn increases the effective field in the dielectric. The increased field causes more deformation, further increasing the effective field, until the thickness of the elastomer drops below a critical limit, allowing for dielectric breakdown to occur. This instability is reduced in elastomers that have *strain stiffening* behavior as the elastomeric chains reach their maximum length and the elastomer stiffens significantly.

Advantages of dielectric elastomer transducers are the simplicity of the system, direct electric drive and facile integration with electrical robotic components, quiet operation, high electro-mechanical efficiency, and performance on par with natural muscles in terms of specific energy and power density.

Disadvantages of the DETs as actuators are the requirement for high voltage, the need for rigid components to maintain prestretch, and the risk of dielectric breakdown. High voltage is a relative term, as most DEAs are driven in the 10–50 V/µm range and voltages higher than 2 kV are needed for appreciable deformation in dielectric films larger than 50 µm.

4.3 Maxwell stress and electro-mechanical performance

The mechanical deformation is a response to the Maxwell pressure due to electrostatic attraction between the compliant capacitors. The formula for the Maxwell pressure is

$$p_{\text{Maxwell}} = \epsilon_0 \epsilon_r \left(\frac{V}{t}\right)^2, \tag{4.1}$$

where

- ϵ_0 is the vacuum permittivity, equal to 8.85×10^{-12} F/m,
- ϵ_r is the dielectric constant of the elastomer, typically in the 1–5 range and unitless,
- V is the applied voltage in Volts, and
- t is the thickness of the dielectric elastomer in meters.

The strain in the elastomer depends on both the Maxwell stress and mechanical properties of the material, more specifically its Young's modulus Y (in units of pressure, or N/m²). A standard assumption we apply in this chapter is that the elastomer is incompressible and has a Poisson's ratio of 0.5. The area strain λ_{area} of the elastomer can then be obtained from

$$p_{Maxwell} = \sigma = \lambda_{area} \times Y, \tag{4.2}$$

where σ is the stress in the elastomer.

To compare the dielectric elastomer actuators with other technologies or natural muscles, it is useful to be able to measure or determine the specific energy density or the volumetric energy density in the system. The units of volumetric energy density are typically reported as J/L or Joules per liter of actuator. For reference, natural muscles operate between 0.4 and 40 J/L and correspondingly around 0.4 and 40 J/kg for specific energy density given the density of muscle being close to 1 kg/L.

From a simple dimensional analysis it is straightforward to see the units of volumetric energy density are the same as those of pressure:

$$1\frac{J}{L} = 1\frac{N \times m}{10^{-3}m^3} = 10^3 \frac{N}{m^2} = 1\,\text{kPa}. \tag{4.3}$$

As shown in Figure 4.2, the total available mechanical energy in an actuator is half the product of the displacement and the blocked force produced. The slope of the line connecting maximum displacement and maximum blocked force is proportional to the material's Young's modulus. Extreme cases such as soft or stiff elastomers are both captured in this view: soft elastomers show large displacement but low force output, whereas the stiffer elastomers have limited displacement but large force outputs.

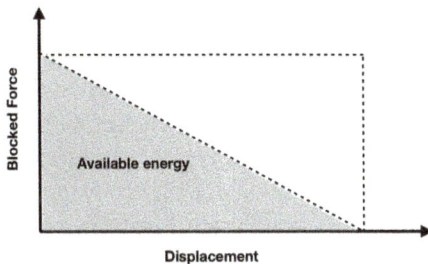

Figure 4.2: Relationship between force and displacement output of a given actuator. The area under the curve corresponds to the available mechanical energy.

Writing the energy equation in terms of the stress and strain, we see that the mechanical energy (e_{mech}) is equal to

$$e_{mech} = \lambda_{area} \times \sigma = \lambda_{area}^2 \times Y. \tag{4.4}$$

Writing this again in terms of material and operational parameters, we see that

$$e_{mech} = \epsilon_0^2 \epsilon_r^2 Y \left(\frac{V}{t}\right)^4. \tag{4.5}$$

For design of dielectric elastomer actuators, we see that the material properties play a significant role: stiffness is directly proportional to mechanical energy. In contrast, mechanical energy depends on the square of the dielectric constant, which means that the constant's impact is more significant. Additionally, in some applications, changing the Young's modulus is not practical, as the overall stiffness of the material may have structural implications for the robot or device. More importantly, the applied electric field has an even higher impact, as the mechanical energy depends on the 4th power of the field. In a simple example, pushing a DEA to twice the electric field means a 16× increase in specific energy, making applied field a critical operational parameter.

4.4 Practical considerations: elastomers and electrodes

Knowing the above physical behavior, sustained efforts have gone into developing improved elastomers by tailoring the composition to maximize response to electrical stress. Similarly, electrodes have been developed for the goal of high electrical conductivity, across large areas, with minimal thickness and minimal impact in terms of increasing the elastomer stiffness. From an integration perspective, the expanding dielectric membrane has minimal robotic utility. Alternative approaches have been developed, making DEAs into bending beams, expanding cylinders, and linearly contracting actuators.

The most established elastomer materials are siloxanes, particularly polydimethlysiloxane (PDMS). These materials have several advantages when being used as dielectric elastomers: they have low viscoelasticity, so they can operate at high frequencies required for applications like haptic communication. Additionally, when PDMS is ablated thermally, it produces silica, SiO_2, as the reaction product, which is an insulator, preventing further damage if dielectric breakdown occurs. The main drawback of PDMS is the low surface tension, which makes some variants difficult to bond to other materials, leading to integration challenges. An additional drawback is the comparatively low dielectric constant, around 3 for most polymer chemistries, and the low breakdown field (30–40 V/µm). Most PDMS compositions are cured thermally, from two-part mixtures, relying on hydrosilylation reactions. Broadly speaking, PDMS can be considered a biocompatible material and has extremely long calendar life, surviving extreme conditions.

A popular option to replace PDMS is a commercially available pressure sensitive adhesive based on an acrylic chemistry. The most widely available variant is VHB from 3M, which typically requires prestretch in the 3 × 3 range biaxially to allow for large deformation. Variants of acrylic chemistries have been developed to address two of the main drawbacks: the need for large prestretch ratios and the high viscoelasticity of VHB. Typically, these acrylic mixtures are cured under ultraviolet light in the absence of oxygen and produce elastomers with higher dielectric constants compared to silicones (4.5 vs. 3). In contrast with PDMS, acrylic materials and their blends with urethane and polybutadiene can be bonded more easily with other materials, easing integration concerns.

The main electrode types are presented in Figure 4.3, following the same principles outlined in Chapter 2 when describing stretchable conductors. Some unique challenges applicable to electrodes for dielectric elastomer transducers are as follows:

- Electrodes need to remain *conductive under large area strains* as large as 1000 %, or 10× the original area. Example 4.3(a): dry dispersed particles show poor performance at large strains. One modification is to use high-aspect-ratio conductors, such as carbon nanotubes or silver nanowires, or graphene nanoribbons, which maintain percolative electrical contacts even at large strains.
- The electrodes should allow for *multiple capacitors to be stacked together*: multi-layered DEAs have larger force responses, proportional to the number of layers. Example 4.3(b): particles dispersed in oil or grease are well suited for single-layer devices but limit how a second layer of elastomer could be bonded to the first.
- The electrodes should have low stiffness compared to the elastomer: if some of the Maxwell pressure is needed for electrode deformation, then less mechanical energy is available for useful work done by the DEA. Example 4.3(c): particles dispersed in an elastomeric matrix (e. g., carbon black dispersed in PDMS) typically have higher modulus than the pure elastomer but achieve strong bonding to the primary elastomer.

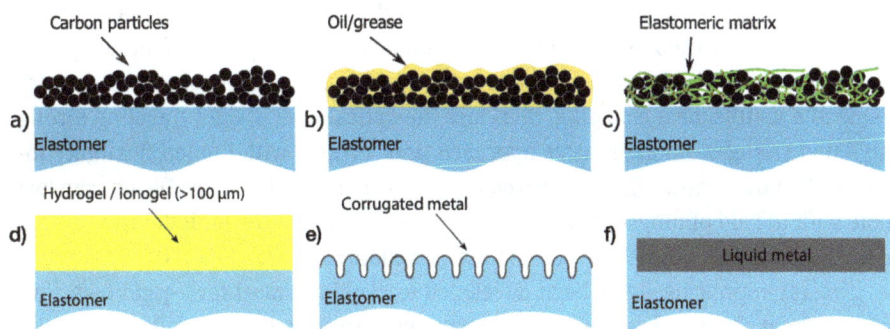

Figure 4.3: Schematics of the most popular electrode types: (a) dry dispersed carbon particles, (b) carbon particles dispersed in oil or grease, (c) carbon particles embedded in an elastomeric matrix, (d) hydrogel or ionogel coatings, (e) metal dispersed onto prestretched elastomer and released to accommodate strain mismatch via corrugation, (f) encapsulated liquid metal.

–   The electrodes should be as thin as possible compared to the elastomer: if the electrode occupies significant space, then the volumetric energy density of the actuator is reduced. Example 4.3(d): layers of liquid–solid composites (e. g., hydrogels or ionogels), where an ionically conductive liquid (e. g., aqueous solution or ionic liquid) is dispersed in a stretchable network. Typically, these gels are made from liquids and cured under ultraviolet light: the surface tension of the freestanding liquids causes large final thickness of the gels.

–   The electrodes should be made at the same strain state the elastomer is in: if the electrode is deposited when the elastomer is under strain, then manufacturing becomes increasingly more complex. Example 4.3(e): metal particles deposited when the elastomer is strained have high, metallic-like conductivity but need to form a corrugated pattern when the elastomer is returned to a zero strain state. For single-layer devices, this approach is relatively easy, but in a multilayer configuration, the complexity of manufacturing is significant.

–   The electrodes should be as light as possible compared to the elastomer: if the electrode has significant weight, then the specific energy density of the actuator is reduced. Example 4.3(f): an encapsulated liquid metal is an excellent conductor. One of the most popular options is an eutectic mixture of gallium and indium, commonly referred to as eGaIn. The density of eGaIn is 6.3 g/mL, so it needs to be used in ultrathin layers to not significantly limit the actuator performance.

4.5 Integration challenges: harnessing mechanical work

The fundamental operation mode of a dielectric elastomer actuator consists of membrane expansion in the plane and contraction within its thickness. This operation mode has limited functionality in terms of producing meaningful work for a mechanical device such as a soft robot. Several applications have been developed, which use the deformation of the membrane to modify the path of light for tunable lenses or laser speckle reducers. Figure 4.4 shows a tunable optic example in which a pocket of liquid operates as a tunable lens. The pocket is compressed by a surrounding DEA supported by a rigid frame and its thickness changing, therefore the optical path of light traveling through the liquid changes. An alternative approach uses a frame with limited flexibility that bends when the central DEA is powered up. Robots made with flexible frames can locomote with the aid of differential friction legs, moving within the limit of viscoelasticity of the elastomer and the RC constant of the compliant capacitor.

Several approaches have been developed to avoid the need for a rigid frame and operate the dielectric elastomer actuators without prestretch. The result is typically a multilayered DEA that has small displacement (< 20 %) but larger force outputs. These devices, which operate without prestretch, can be more readily assembled into devices that produce mechanical work valuable for soft robotics. Figure 4.5 shows the simplest deformation mode: bending when adhered to a strain limiting layer, which is capable

(a) Driving "off"   **(b)** Driving "on"

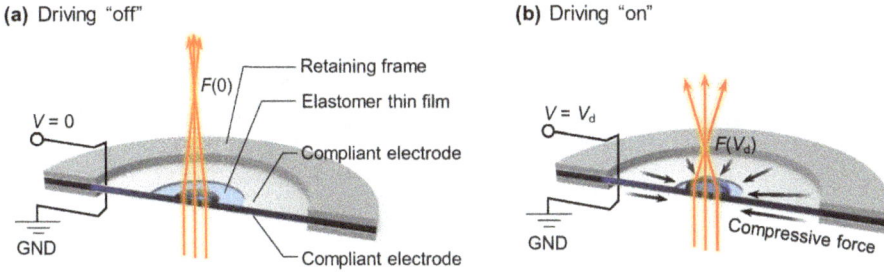

Figure 4.4: Research example of a tunable lens deformed by a dielectric elastomer actuator surrounding the central liquid pocket. Reproduced with permission from "Monolithic focus-tunable lens technology enabled by disk-type dielectric-elastomer actuators".

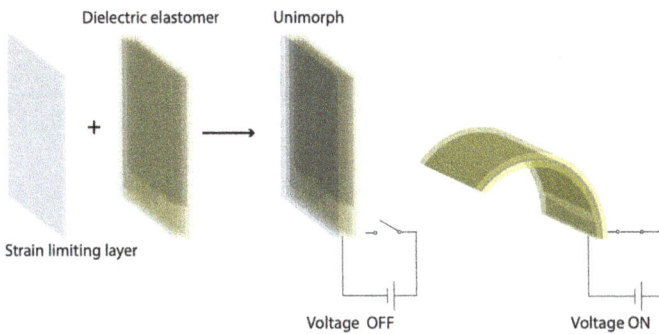

Figure 4.5: Schematics of the construction of a bending DEA: a strain limiting layer is adhered to an actuator. When powered up, the actuator bends toward the strain-limiting layer, operating as a unimorph.

of bending but not stretching. A single bending actuator can be used in a broad range of configurations: a gripper finger, a crawling inchworm-inspired robot, and a high-deformation-rate jumping or hopping locomotion.

Logically, the single actuator can be attached to another identical DEA to produce a bimorph system capable of bending in two directions. Care must be taken to match the bending performance of the two actuators to produce a symmetric system. A method of assembly is shown in Figure 4.6, and these types of actuators can be used in more complex robots, such as fish-inspired robots that deform their bodies to swim in water.

While bending actuators are relatively useful, the easiest actuator to integrate is something that moves linearly, either in expansion or contraction. Figure 4.7 shows two methods to produce linear actuators from single or multilayer DEAs. Part **A** shows that rolling a long actuator along one axis produces a cylindrical actuator. After connections are made, when the cylinder is powered, it expands radially and axially, overall showing linear expansion relative to its initial length. Radial expansion can be reduced through reinforcements, such that primarily the cylinder increases in height with min-

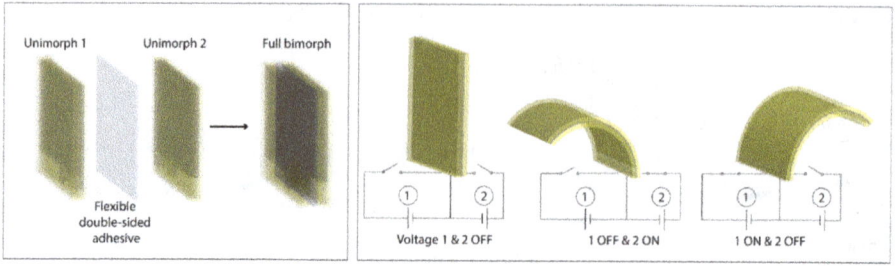

Figure 4.6: Schematics of the construction of a bimorph DEA made from two separate benders, attached to a central double sided adhesive.

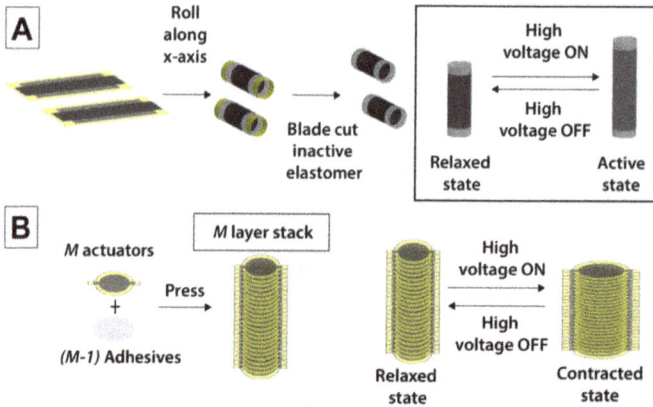

Figure 4.7: Schematics of the two methods of producing linear deformation from DEAs: (a) rolling a long actuator into an expanding cylinder and (b) stacking multiple actuators to produce a linear contractor. Alternative methods of making linear contractors use folding of single actuators on themselves, which is a specific case of stacking multilayered actuators.

imal changes to radius. Part **B** shows that multiple membranes can be stacked on top of each other to produce a high-aspect-ratio actuator. When powered, each membrane deforms in response to Maxwell pressure, leading to overall contraction in the long axis of the actuator. Both variants have had multiple robotic applications, from pipe inspection to haptic interfaces, flapping wing robots, and artificial limbs.

4.6 Alternative approaches: HASELs and generators

A variant of DEAs can be made by replacing the dielectric elastomer with a dielectric fluid. These types of actuators are called **HASELs**, which stands for Hydraulically Amplified Self-Healing Electrostatic Actuators. The fundamental principle is the same as in

DEAs: an applied electric field causes Maxwell pressure to develop between the electrodes. This Maxwell pressure causes the liquid to redistribute within the structure, causing bending, expansion, or contraction, depending on how the HASEL is assembled. A significant advantage over DEAs is the ability of the liquid to fill in gaps where dielectric breakdown occurs, giving the system its "self-healing" ability. Some drawbacks include the risk of leakage of the liquid dielectric and the requirement for high voltage, typically higher than DEAs because of the need to overcome capillary forces in the liquid dielectric. Another challenge is at the power electronics level: unlike DEAs, HASELs need to be driven by fields of alternating polarity to avoid species migration in the dielectric fluid. This requirement increases the size and complexity of the electronics required to make the system autonomous.

Dielectric elastomer generators are a variation of how the transducer operates to convert some mechanical energy applied to the capacitor into electrical work. Figure 4.8 shows a schematic of the operation mode, specifically focusing on the mechanical strain state of the system and the field across the capacitor. The most common proposed application of DEGs is wave energy harvesting, setting up the system to be deformed by waves in the ocean at the water surface.

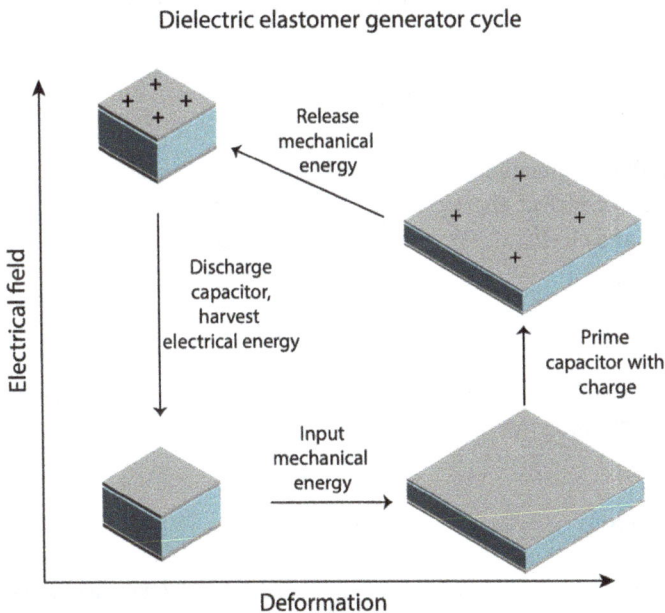

Figure 4.8: Schematic of the steps in the cyclic operation of a dielectric elastomer generator.

The steps can be summarized as follows:
- The DEG starts at a 0 % strain state and completely discharged.
- As the DEG is strained, mechanical energy is stored in the compliant capacitor. The mechanical energy is

$$e_{mech_2} = \lambda^2 \times Y.$$

- A priming charge is applied in the expanded state at a relatively small voltage. The input electrical energy is

$$e_{elec_3} = \frac{1}{2}C_3 \times V_3^2.$$

- When the mechanical strain is released, the DEG returns to an intermediary strain state, which greatly reduces its capacitance. Because of charge conservation, the voltage across the capacitor must increase:

$$Q_3 = Q_4 = C_3 \times V_3 = C_4 \times V_4.$$

The increase in voltage is due to the conversion of some of the mechanical energy into electrical energy:

$$e_{elec_4} = \frac{1}{2}C_4 \times V_4^2.$$

- The high voltage at the intermediary state can be discharged into auxiliary power electronics to harvest electrical energy. The DEG is returned to its original zero strain state. An mechano-electrical energy efficiency can be calculated as

$$\eta = \frac{e_{elec_4} - e_{elec_3}}{e_{mech_2}}.$$

The priming energy can be significant in small DEGs, which impacts the scale of construction for effective systems.

4.7 Example laboratory session

In this laboratory session, students will make and test dielectric elastomer actuators. We will have two separate sets of operations: making devices (prestretching, masking, and painting electrodes) and testing (capacitance and resistance measurements) and actuation with high-voltage amplifiers. All the components required for assembly are listed in Figure 4.9 and below.

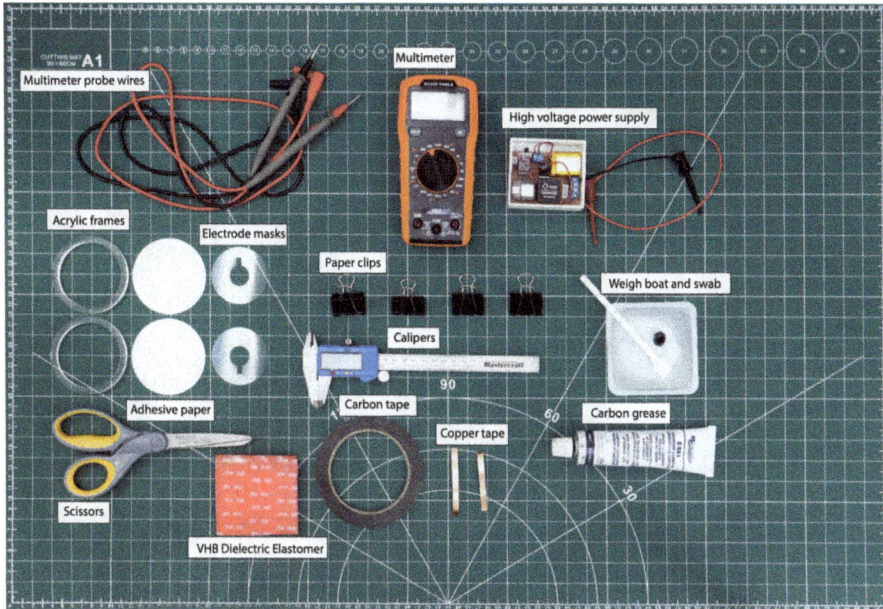

Figure 4.9: Required components for the assembly of prestretched dielectric elastomer actuators.

The DEA fabrication station will include:
-   VHB dielectric elastomer,
-   adhesive paper for affixing onto frames,
-   acrylic frames made via laser cutting,
-   masks for electrodes,
-   carbon grease to be painted as electrodes, and
-   conductive tape (carbon or copper) or wires to make connections.

The DEA evaluation station will include:
-   LCR meter for capacitance measurement,
-   multimeter for resistance measurement, and
-   connectors to the device.

The DEA testing station will include:
-   high-voltage amplifier and
-   rulers and cameras to track displacement.

The fabrication procedure uses commercial VHB 4910 elastomer as the stretchable dielectric. The students should follow the visual instruction in Figures 4.10, 4.11, 4.12, 4.13.

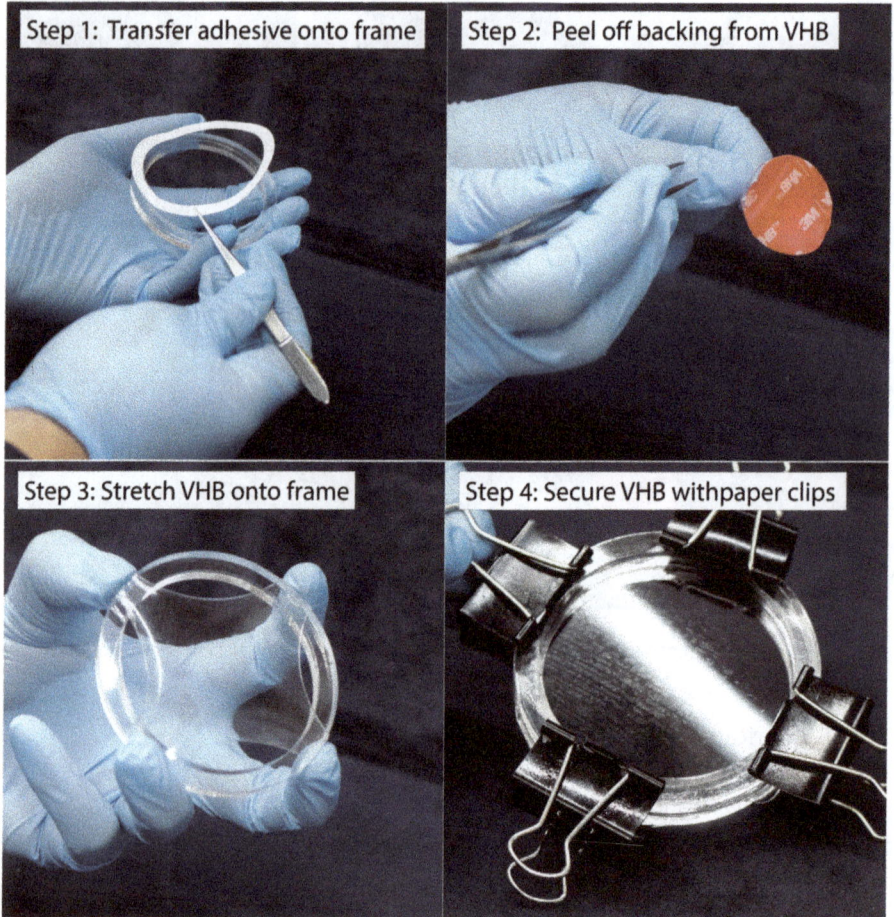

Figure 4.10: Steps 1–4 from assembly procedure.

- Step 1: Remove the paper backing from the adhesive and affix it onto the acrylic frame. Repeat for the second frame.
- Not pictured: Mark 4 corners of a 2 cm by 2 cm square on the VHB elastomer before stretching.
- Step 2: Remove the backing from VHB elastomer.
- Step 3: Stretch the elastomer to an area large enough to stick to the frame.
- Not pictured: Use a frame with sticky tape to secure the elastomer and fix the strain state.
- Not pictured: Attach a frame on the opposite side.
- Step 4: Secure frame to each other with paper clips.
- Step 5: Place masks for electrodes on one side.

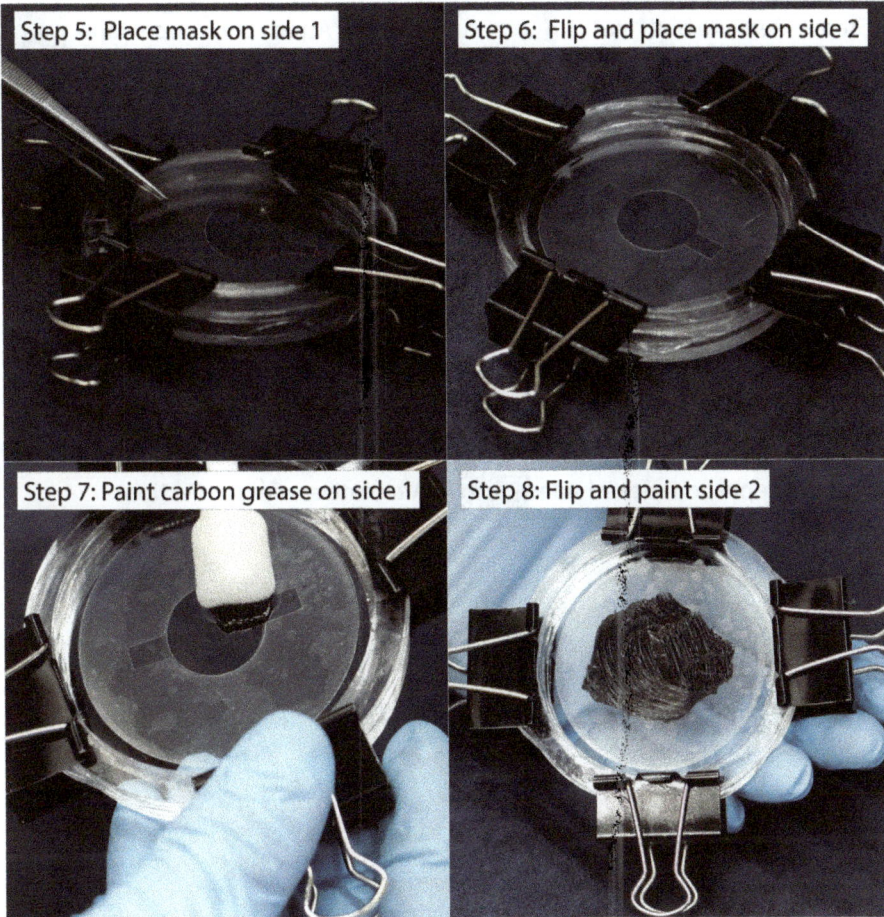

Figure 4.11: Steps 5–8 from assembly procedure.

- Step 6: Place mask on elastomer on opposite side. Make sure to have the electrode connections pointing away from each other to avoid electrical shorts.
- Step 7: Paint carbon grease as electrodes on one side of the elastomer.
- Step 8: Flip the frame and paint the second side.
- Step 9: Remove the masks.
- Steps 10 and 11: Attach carbon tape (or copper, or wires) to make electrical connections.
- Move to the next station, measure capacitance and resistance, and record them.

To measure mechanical strain in response to Maxwell stress, the students need to be able to apply a high voltage. For this experiment, we will use the DEAs made previously

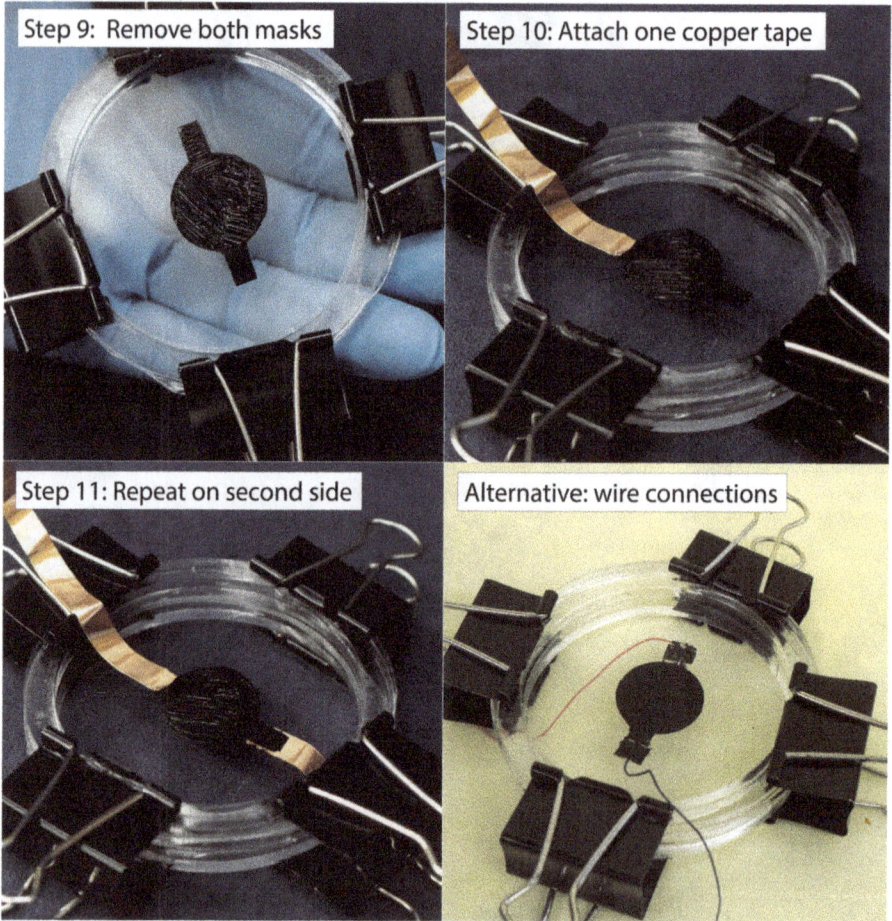

Figure 4.12: Steps 9–11 from assembly procedure, with an alternative connection method: wires instead of copper tape.

and test them as biaxial expanders. By keeping track of the diameter of the electrode as the device expands you can measure axial strain and plot it as a function of the applied voltage.

1. Connect the DEAs to a high voltage amplifier.
2. Measure the initial diameter d_0.
3. Apply voltage in 0.5 kV increments and measure the diameter at that state. Take photos of the DEA from above with a ruler in the background to record diameter.
4. Increase voltage until dielectric breakdown.
5. Record values for each type of electrode and pre-stretch ratio.

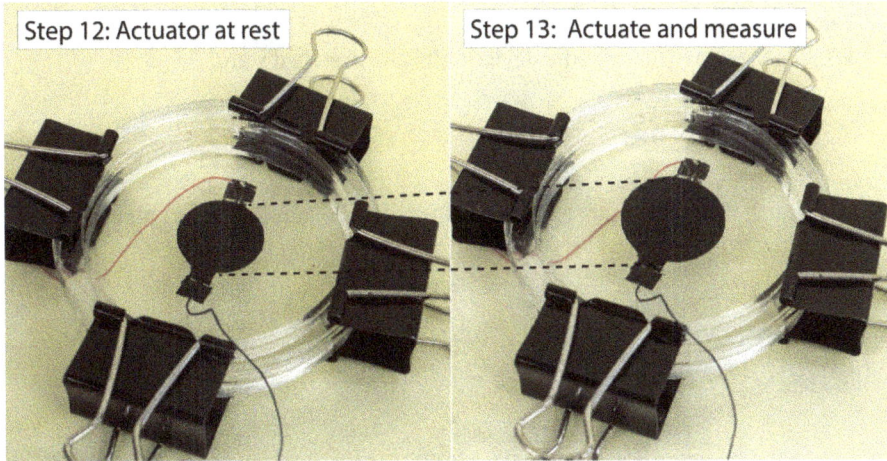

Figure 4.13: Steps 12–13: actuator on and off.

4.8 Example problems

Problem 1. To better understand the potential of energy harvesting, let us consider a stretchable capacitor with an area of 100 cm^2 and a thickness of 300 μm. Assume that the dielectric material is VHB 4910 and has a dielectric constant of $\epsilon = 4.5$ and a modulus of approximately $Y = 200$ kPa. Assume that the electrodes are ideal, add no appreciable stiffness or thickness to the device, and maintain conductivity when strained 10× their original length. Recall the capacitance formula

$$C = \epsilon\epsilon_0 \frac{A}{d},$$

where C is the capacitance, A is the area, d is the dielectric thickness, ϵ is the relative permittivity, and ϵ_0 is the permittivity of free space (8.85×10^{-12} m^{-3} kg^{-1} s^4 A^2).

Also note that the capacitance depends on the charge in the capacitor and the voltage applied to it:

$$C = \frac{Q}{V},$$

where C is the capacitance, Q is the charge stored in the capacitor, and V is the voltage applied.

Lastly, recall that the energy stored in a capacitor is

$$E = \frac{1}{2}CV^2.$$

Calculate the following values:

1. What is the capacitance of the device in its nonstretched state?

$$C_1 = \epsilon\epsilon_0 \frac{A}{d} = 4.5 \times 8.85 \times 10^{-12} \, m^{-3} \, kg^{-1} \, s^4 \, A^2 \times \frac{10^2 \, cm^2}{300 \, \mu m} \times \frac{1 \, m^2}{10^4 \, cm^2} \, \frac{10^6 \, \mu m}{1 \, m} = 1.32 \, nF.$$

2. How much does capacitance increase when stretched to 10× the original area?

$$C_1 = \epsilon\epsilon_0 \frac{A_1}{d_1},$$

but

$$C_2 = \epsilon\epsilon_0 \frac{A_2}{d_2} = \epsilon\epsilon_0 \frac{10A_2}{\frac{d_2}{10}} = 100C_1 = 132 \, nF.$$

3. How much mechanical energy is stored in the stretched elastomer before priming it with charge?

The volumetric energy density can be found from stress and strain:

$$e_{mech_2} = \lambda^2 \times Y = 10^2 \times 200 \, kPa = 20 \, MPa = 20 \, kJ/L.$$

The volume of the elastomer is

$$V_{volume} = A \times d = 100 \, cm^2 \times 300 \, \mu m = 3 \, mL,$$

leading to the total energy

$$E_{mech_2} = V_{vol} \times e_{mech_2} = 20 \, kJ/L \times 3 \, mL = 60 \, J.$$

4. Assume that the stretched capacitor is primed to a voltage of 100 V. What is the amount of charge stored in the capacitor?

$$Q_3 = C_3 V_3 = 132 \, nF \times 100 \, V = 13.2 \, \mu C.$$

5. Keeping in mind that charge has to be conserved, what is the voltage when the applied mechanical stress is released, and the capacitor returns to its original size?

$$Q_2 = C_1 V_1 = Q_2 = C_2 V_2,$$

$$V_2 = \frac{C_1}{C_2} V_1 = \frac{132}{1.32} 100 = 10 \, kV.$$

6. How much electrical energy is stored in the capacitor at the two states? Report your answer in Joules.

In the stretched state,

$$E_{elec_3} = \frac{1}{2}C_3V_3^2 = \frac{1}{2}132\,\text{nF} \times 100^2\,\text{V}^2 = 0.66\,\text{mJ}.$$

In the relaxed state,

$$E_{elec_4} = \frac{1}{2}C_4V_4^2 = \frac{1}{2}1.32\,\text{nF} \times 10^8\,\text{V}^2 = 66\,\text{mJ}.$$

7. What is the electro-mechanical conversion efficiency? That would be the ratio of the difference between the values of items 6 and 3.

$$\eta = \frac{E_{elec_4} - E_{elec_3}}{E_{mech_2}} = \frac{0.066 - 0.00066}{60} = 0.11\,\%.$$

Problem 2. A dielectric elastomer actuator is designed as a 1-nF capacitor with an internal resistance of 100 kΩ. The DEA is actuated by applying a voltage signal, and its deformation is assumed to be instantaneous with the applied electric field (i. e., no viscoelastic effects). However, due to the RC time constant of the system, the voltage across the DEA does not change instantly, affecting the response time. Find a way to estimate the charging time required for the DEA to reach actuation voltage. In other words, what is the response speed of the actuator based on its electrical characteristics? Would this be suitable for a haptics application where the actuator must operate at 200 Hz?

Solution. The DEA behaves like an RC circuit, where the charging behavior follows

$$V(t) = V_0(1 - e^{-t/\tau}),$$

where V_0 is the applied actuation voltage, $\tau = RC$ is the RC time constant, $R = 100\,\text{k}\Omega$ is the resistance, and $C = 1\,\text{nF}$ is the capacitance.

The time constant is

$$\tau = (100 \times 10^3) \times (1 \times 10^{-9}) = 0.1\,\text{ms}.$$

This means that the voltage reaches approximately
- 63 % of V_0 at $t = \tau = 0.1\,\text{ms}$,
- 95 % of V_0 at $t \approx 3\tau = 0.3\,\text{ms}$, and
- 99 % of V_0 at $t \approx 5\tau = 0.5\,\text{ms}$.

Since the DEA deforms fast enough to charge and discharge in 1 ms, it is fast enough for the haptic operation where each cycle would be 5-ms long.

Problem 3. A dielectric elastomer actuator is modified to include a liquid crystal elastomer layer, where the stiffness in the oriented direction is 100× larger than in the orthogonal direction to it. Assume that the elastomer is a rectangle (5 cm by 1 cm), and the layer is 50-μm thick. The stiffness in the "soft" direction is 200 kPa. Find the strain in each dimension when the device is actuated at 5 kV. Assume that the dielectric constant of the material is 5.

Problem 4. A dielectric elastomer capacitive sensor is operated uniaxially. The electrodes are sputter-coated gold, which has been deposited when the material had been prestretched uniaxially. As a result, when deformed uniaxially again, the dimensions of the capacitor change from $L_0 \times T_0 \times W_0$ to $L_x T_x W_0$. Keep in mind that this material has a Poisson's ratio of 0.5. Draw a schematic of how the system operates between the two strain states. Determine the initial and final capacitances of the sensor if it is stretched from an initial length of 5 cm to a final length of 25 cm. The initial thickness is 1 mm, the initial width is 2 cm, and the elastomer dielectric constant is 2.

Bibliography

Bong Je Park et al. "Monolithic focus-tunable lens technology enabled by disk-type dielectric-elastomer actuators". In: *Scientific Reports* 10.1 (2020), pp. 16937.

5 Thermal powered machines

Contents

Machines powered by thermal energy have been widely studied because of the availability of phase-changing materials that produce deformation or force. This chapter focuses on the three types of phase changes: between crystallographic structures in shape memory alloys, between liquid and vapor in encapsulated fluids, and between different internal structures, such as nematic to isotropic in liquid crystal elastomers.

5.1 Learning objectives

The main concept to be presented and understood is how phase transitions require an input of energy, proportional to the mass of material, and its properties. Three material classes are discussed: shape memory alloys (which are much stiffer than typical elastomers and so offer large forces), encapsulated liquids in elastomers (which undergo a phase transition to a gas and cause expansion), and liquid crystal elastomers (which undergo a change in structure that leads to deformation). For all these systems, the two key challenges are integration with the rest of the soft machine and having a means to deliver thermal energy rapidly.

5.2 Background and principles

The concept of shape memory has been known since 1932, when Swedish chemist Arne Ölander observed it in gold–cadmium alloys. Several decades later, William J. Buehler [1] along with Frederick E. Wang, working at the Naval Ordnance Laboratory (NOL), discovered and characterized a nickel–titanium alloy, which showed that it could be deformed and heated to return to its original shape. The common name for the material, *nitinol*, comes from the abbreviations of the two elements that form the alloy and the location of the discovery (NOL). Meanwhile, liquid crystal elastomers were first theorized by Pierre-Gilles De Gennes in 1975 and synthesized by multiple researchers in the follow-

https://doi.org/10.1515/9783111069418-005

ing years. Lastly, phase change materials encapsulated in elastomers are a more recent invention, emerging from Prof. Hod Lipson's group at Columbia University during the recent boom of interest in Soft Robotics.

Figure 5.1 shows the change in temperature for a specific amount of water when heated at a constant rate from below its melting point (0°) to above its boiling point (100°) at 1 atmosphere of pressure. It is immediately clear that when the system is undergoing a phase transition, the temperature is constant up to the point where the entire material changes phases. Meanwhile, in the single phase regions, the change in temperature is proportional to the change in heat added to the system. Mathematically, this can be expressed as

$$\Delta Q_1 = mC_w \Delta T_1, \tag{5.1}$$

where
- ΔQ_1 is amount of heat added to the liquid water,
- m is the mass of the water,
- C_w is the specific heat of the liquid water, a constant, and
- ΔT_1 is the change in temperature for the liquid water.

Meanwhile, for the phase change region,

$$\Delta Q_2 = m\Delta H_{vap}, \tag{5.2}$$

where ΔH_{vap} is the specific enthalpy of vaporization for the water, expressed in J/g.

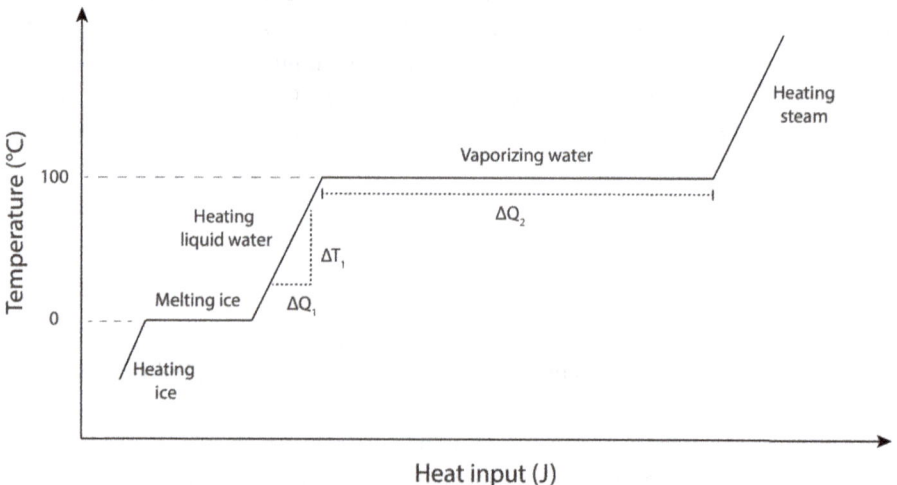

Figure 5.1: Diagram of the change in temperature of water when heated at a constant rate from below the freezing point to above its boiling point.

The change in phase is usually accompanied by a change in volume, which can be leveraged to produce deformation or force in a soft machine. The challenge for all these phase-changing materials is to design and integrate the active matter so that it can easily be powered, and its output is maximized relative to the goal of the soft machine. The next sections cover the fundamentals of three different types of materials, all of which undergo valuable phase changes. The amount of thermal energy delivered via a phase transition described by equation (3.2) will always be the maximum available energy for mechanical deformation.

5.3 Shape memory alloys

The most commonly used shape memory alloy (SMA) is nitinol, which has unusual properties due a reversible solid-state phase transformation between two crystal structures, austenite (stable at high temperatures) and martensite (stable at low temperatures). This change is also called a martensitic transformation. From a mechanistic perspective, at high temperatures, nitinol has interpenetrating simple cubic structure referred to as *austenite*. At low temperatures, nitinol transforms to a more complicated monoclinic crystal structure known as *martensite*. The switch from austenite to martensite is both reversible and instantaneous in both directions.

Figure 5.2 shows a thermo-mechanical cycle at both the atomic and macroscales. The starting point may be martensite, at low temperature and low strain, where its crystal structure is monoclinic. In this state, martensite has can undergo deformation

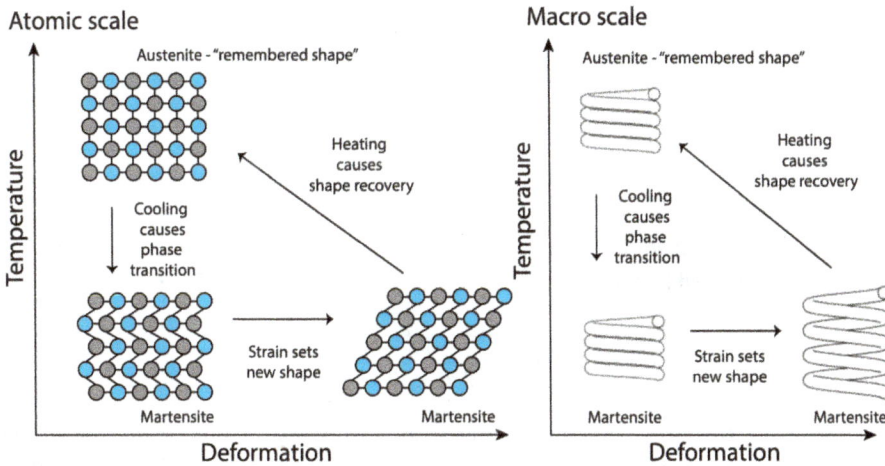

Figure 5.2: Comparison of how a shape memory alloy transitions between states and recovers shape at the atomic and macro scales. The austenite phase is cooled and changes into martensite. The martensite can be deformed and retains the new shape at low temperature. When taken to the high temperature, the material returns to its original, "remembered" shape and converts back into austenite.

of 6–8 % without breaking any bonds. This process is called twinning and consists of the rearrangement of atomic planes without causing slip, or permanent deformation. When heat is applied, the material seeks to return to the more stable original austenitic structure, and the strain is reversed, returning to the original shape. Alone, SMAs do not have the ability to power a soft machine in a reversible fashion and need to be paired up with a soft material that returns them to the high-strain state when cooled. The driving force for such a machine is heat applied that causes SMA contraction, followed by elastomer-driven expansion when cooled. A common challenge in powering SMA-based soft machines is integration of small, precise heaters, usually resistive ones, that can controllably trigger deformation.

To give the reader some design boundaries, let us consider an SMA that has to be integrated in a soft structure for elastic return to high strain and compare the amounts of materials needed. For the soft elastomer to push the SMA back to its high-strain state, it must store the same amount of energy the SMA requires when contracting,

$$E_{SMA} = E_{elastomer}. \tag{5.3}$$

The mechanical energy is a product of stress σ, strain λ, and the amount of material: SMAs undergo about 3 % strain at a 75 GPA modulus:

$$\frac{1}{2}m_{SMA} \times \rho_{SMA} \times \sigma_{SMA} \times \lambda_{SMA} = \frac{1}{2}m_{elastomer} \times \rho_{elastomer} \times \sigma_{elastomer} \times \lambda_{elastomer}, \tag{5.4}$$

where
- $\rho_{SMA} = 6.5\,g/mL$,
- $\rho_{elastomer} = 1\,g/mL$,
- $\sigma_{SMA} = Y_{SMA}\lambda_{SMA}$,
- $\sigma_{elastomer} = Y_{elastomer}\lambda_{elastomer}$,
- $Y_{SMA} = 30\,GPa$ for martensite,
- $Y_{elastomer} = 1\,MPa$, and
- $\lambda_{SMA} = 3\,\%$.

This leads to a design equation linking the mass of the two components to the amount of strain experienced by the elastomer:

$$\frac{m_{SMA}}{m_{elastomer}} = \frac{1}{6.5}\frac{10^6}{75 \times 10^9}\frac{\lambda^2_{elastomer}}{1.03^2}. \tag{5.5}$$

Practically, the large modulus and low strain of the SMA imply a requirement for large amount of elastomer to compensate and match the energy stored in the SMA. One design trick, widely employed, is to give the SMA a tortuous, or serpentine path, leading to a much larger strain at the end points, as shown in Figure 5.3. This approach increases the total strain in the structure and the equivalent strain in the elastomer, allowing for less elastomer to be needed and simplifying robot design.

Linear design   Serpentine design

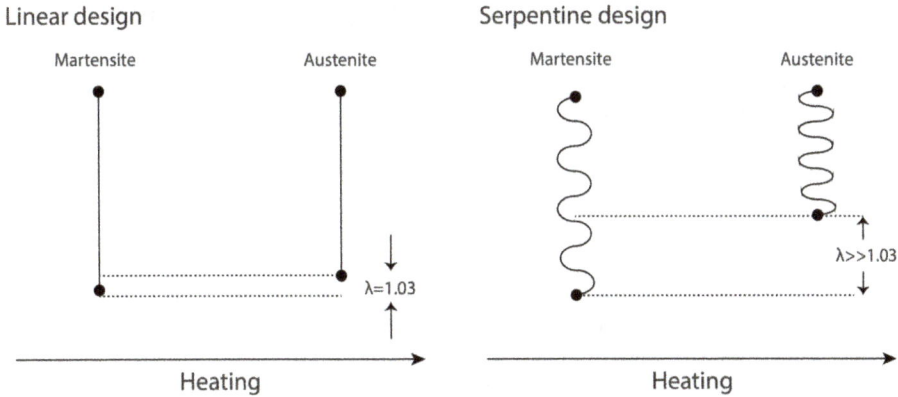

Figure 5.3: Comparison of how a linear SMA undergoes small strain, around 3 %, whereas a serpentine design shows much larger strain between the end points when heated because the actual path of the SMA is longer in the serpentine design.

Superelasticity (SE), sometimes termed "pseudo-elasticity" or "pseudo-plasticity", occurs without any change in temperature. SE takes place at temperatures above As (temperature at which the austenitic phase begins to form), although usually only slightly above, where the austenitic phase is thermodynamically more stable of the two, although not very much. When a mechanical strain is imposed, this can stimulate the transformation of austenite to martensite, sometimes termed "stress-induced martensite". The associated shear of local regions accommodates the imposed macroscopic shape change, whereas the lower strain energy component ensures that the overall free energy is now lower than it would be if the austenitic phase were still predominant. This behavior is not leveraged in soft machines but useful in medical devices where large deformation of metals is needed, for example, transcatheter pulmonary valves.

5.4 Encapsulated phase-changing fluids

A different type of thermal actuator uses phase-changing fluids, encapsulated in a supporting elastomer. Initially, at low temperature, the fluid is a liquid that when heated, undergoes a phase transition to a vapor and expands significantly. If neither the fluid nor the vapor permeates through the elastomeric matrix, then the entire composite expands and applies force to its surroundings if constrained. Figure 5.4 shows a diagram of a linear actuator, made from one of these composites, with a single resistive heating element running at through its core.

We can estimate the amount of force or strain in such an actuator from the change in volume during the vaporization phase. The force on the elastomer (f_e) must be balanced when the vapor expands (f_v):

$$f_e = f_v. \tag{5.6}$$

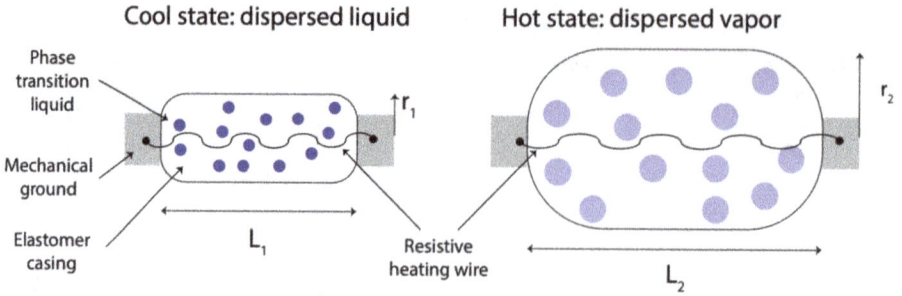

Figure 5.4: Schematic of an elastomer encapsulating a phase-changing fluid. The system is powered by a serpentine resistive heating element running along the length of the elastomer. When the actuator is cool, the dispersed material is a liquid. When heated, the liquid evaporates but cannot permeate through the elastomer and causes the entire system to expand.

Alternatively, this can be written as

$$\lambda_e Y_e = p \Delta V. \tag{5.7}$$

For a common literature example, ethanol expands roughly 19× its original size when evaporating. When encapsulated in Ecoflex 00-50, an elastomer with a modulus of 200 kPa, and starting at 1 atm, the expansion should be at most 9.5× to account for the impact of modulus. Laboratory experiments confirm a 9.15× maximum expansion, proving that the simple approximation relatively well captures the behavior of the system.

Multiple practical considerations must be taken into account when making and operating these types of devices:

1. The evaporation temperature of the liquid must be in the range of interest for the application. Common demonstrations use ethanol as the working fluid, with a boiling point of 78.4 °C.

2. The liquid must not permeate through the elastomer, even when heated to a vapor phase. Literature examples use ethanol and silicone elastomers, typically two-part mixtures that react via a catalyst-driven hydrosilylation.

3. The liquid must not act as a poison toward the uncured elastomer. Many fluids poison the catalyst and drastically increase curing time for the elastomer.

4. The components must be relatively safe for the desired application. For example, ethanol escaping the actuator in a wearable garment is relatively safe, as it can permeate out of the active area. In contrast, if the actuator was used inside the human body, then escaping ethanol would react with the surrounding anatomy and could potentially be harmful.

5. The system must contain a method to deliver heat locally to trigger the phase change. Resistive heating wires, such as NiChrome, are commonly used in a coiled configuration to have the ability to expand along the actuator.

6. Cooling the actuator to return to its original shape is a more significant concern than in shape memory alloys. In SMAs the actuator has nearly the same volume and the same heat transfer coefficient in both states. In phase-changing fluid elastomers, the vapor has lower heat transfer coefficient, the entire composite has larger volume, and both factors slow the return to the original size.
7. These actuators expand in all directions, which causes integration challenges in producing valuable motion for a soft machine. Lessons have been learned from fiber-reinforced actuators, and even early literature examples used McKibben-style reinforcements to make phase-changing elastomer actuators into contracting muscles.

For any thermal actuator, a **thermo-mechanical efficiency** can be calculated as the ratio of mechanical energy produced from the thermal input. The amount of mechanical energy delivered by the actuator uses the same relationship between stress, strain, and Young's modulus as before and is written as:

$$e_m = \frac{1}{2}\lambda_m \sigma_m = \frac{1}{2}\lambda_m^2 Y_m. \tag{5.8}$$

The amount of thermal energy needed to vaporize the working fluid is the sum of the energy needed to heat it up to the phase transition temperature and the amount of heat required to complete the phase transition:

$$e_t = m_l C_l \Delta T + m_l \Delta H_{\text{vap}}, \tag{5.9}$$

where
- e_t is the total thermal energy required,
- m_l is the mass of the phase-changing fluid,
- C_l is the specific heat of the fluid in its liquid form below the phase transition temperature,
- ΔT is the change in temperature from initial T_0 to the vaporization temperature T_{vap}, and
- ΔH_{vap} is the vaporization enthalpy (in J/kg) for the working fluid.

However, delivering thermal energy directly tends to be both imprecise and inefficient. Instead, resistive heating can be used to locally heat the area around the heating element, typically a NiChrome wire. The amount of electrical energy delivered is found from Ohm's law:

$$e_e = \eta_{\text{et}} t I V = \eta_{\text{et}} t I^2 R, \tag{5.10}$$

where
- e_e is the total electrical energy delivered,
- η_{et} is the electro-thermal energy conversion efficiency, a number that can be estimated to better model the system,

- t is the total heating time,
- I is the current applied to the resistive wire,
- V is the voltage across to the resistive wire, and
- R is the electrical resistance of the wire.

The total energy efficiency can be written as the ratio of mechanical output to electrical input as

$$\eta_{em} = \frac{\frac{1}{2}\lambda_m^2 Y_m}{\eta_{et} t I^2 R}. \tag{5.11}$$

Written in this form, the equation shows the operator of the device how to control a desirable output (e. g., speed of deformation through t or amount of deformation through λ_m) through both design (R, Y_m, and η_{em}) and operating conditions (I). Broadly, the electro-thermal efficiency aims to capture variations in the design of the composite, for example, varying the liquid-to-elastomer ratio. The change in thermal efficiency for each composite can be determined at different ratios experimentally and then used in the modeling of the actuator.

Figure 5.5 shows a comparison between multiple actuator types, including those discussed in this book so far. The phase changing ethanol containing elastomer is labeled as *Soft Actuator* in the figure. *PAMs* are pneumatic air muscles, or McKibben-style actuators, *DEAs* are dielectric elastomer actuators, etc. *IPMCs*, also known as ionic polymer metal composites, will be discussed in the chapter on *Advanced Topics*. Overall, thermal

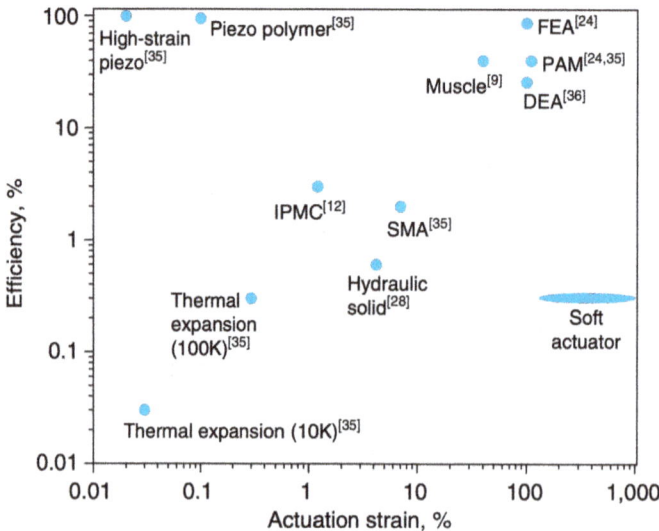

Figure 5.5: Comparison between actuator types in terms of actuator efficiency and strain. The phase changing actuator containing ethanol is labeled *Soft actuator* (Miriyev, Stack, and Lipson, "Soft material for soft actuators").

systems are fairly low in efficiency but offer other advantages such as ease of integration and control through resistive elements. Here we can directly compare the phase-changing system and shape memory alloys: where SMAs are more efficient, they offer significantly less strain.

5.5 Liquid crystal elastomers

Liquid crystal elastomer (LCE) actuators are a variant of thermally driven actuators, where the addition of heat causes a change from an ordered state (i. e., smectic) to a disordered state (i. e., isotropic). The liquid crystal segments that drive the formation of an ordered state are called mesogens, cross-linked with polymer segments that provide the soft support for the composite. Figure 5.6 shows operation of an LCE that contracts when heated to accommodate the LCE structure transitioning from a smectic phase to an isotropic phase.

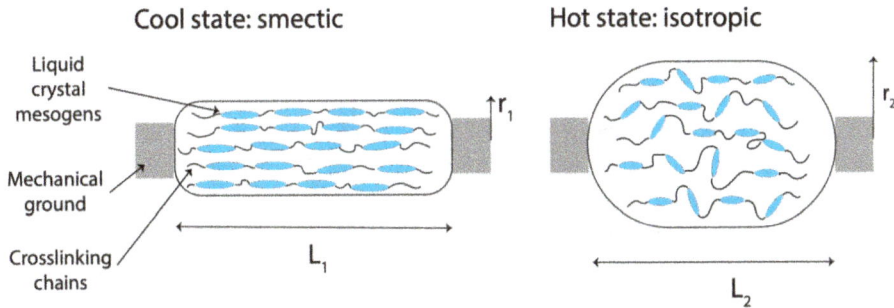

Cool state: smectic   **Hot state: isotropic**

Liquid crystal mesogens

Mechanical ground

Crosslinking chains

r_1 r_2

L_1 L_2

Figure 5.6: Operation of a liquid crystal elastomer actuator, where the addition of heat causes the polymers to reorient from a smectic state to an isotropic state, causing linear contraction in the process.

The transition between states in an LCE is more gradual compared to shape memory alloys. As heat is added, the mesogens begin to realign, converting from the smectic phase to a nematic phase, before reaching the isotropic phase where the mesogens are randomly oriented. Figure 5.6 shows the transition and examples of how the mesogens are incoporated into the LCE structure. Typically, mesogens are rigid segments, and including them in the main chain raises the Young's modulus of the elastomer. Alternatively, connecting the mesogens as side chains manages to keep the material modulus low while retaining the thermal-response ability needed for specific applications. The concepts are illustrated in Figure 5.7, with the mesogens represented as blue ellipses.

LCEs have become a popular approach to solve soft robotic challenges because of the versatility of the system from building blocks to processing methods. In terms of building blocks, multiple groups have been demonstrated to work as mesogens linkers, including vinyl, diacrylates, dihydroxyl, epoxylic, as well as oxetane and dioxetane. The

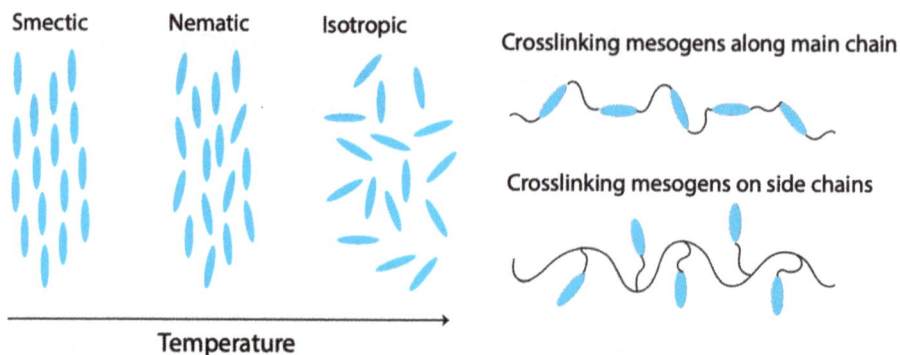

Figure 5.7: Left side: different potential orientation of liquid crystal mesogens, from oriented smectic to disordered isotropic. Right side: example of how mesogens are incorporated into the LCE, either as main or side chain components.

types of reactions that produce LCEs are hydrosilylation, step-growth polymerization, free-radical polymerization, and cationic photopolymerization. Click reactions have also been adapted, including thiol-ene, thiol-yne, thiol-Michael, and aza-Michael. The broad portfolio of reagents and reaction types enables extremely creative approaches to build LCEs and power them externally.

There are multiple methods of assembling the LCEs with the overarching goal of achieving reversible actuation without the need for a preload as well as control of local mechanical responses in both scale and direction. The most common method of orienting the LCEs is mechanical alignment, where the chains are partially cross-linked before mechanical stress is applied and the chains are oriented. Once the mesogens are aligned, the cross-linking process is completed, and the orientation is set. This process produces actuators with uniaxial mesogen orientation capable of significant axial contraction and extension due to the collective mesogen movement throughout the actuator. Other approaches include field-assisted alignment, using the difference in properties between the mesogens and the rest of the chain to align the structure before setting it in place via cross-linking. However, the most research interest in the past years has been in additive manufacturing techniques, in particular 3D and 4D printing, which produces highly complex geometries, including ones with precise voxel control.

5.6 Alternative mesogens: optical and humidity responsive

The liquid crystal elastomer structure is extremely versatile and not limited to responding to thermal stimuli. Figure 5.8 shows a literature example by Lugger et al., "4D printing of supramolecular liquid crystal elastomer actuators fueled by light", of a material capable of responding to both thermal and light signals as a function of the mesogens incorporated in it. The structures are produced via 4D printing, which gives the operator freedom in designing and assembling the structure via direct ink writing.

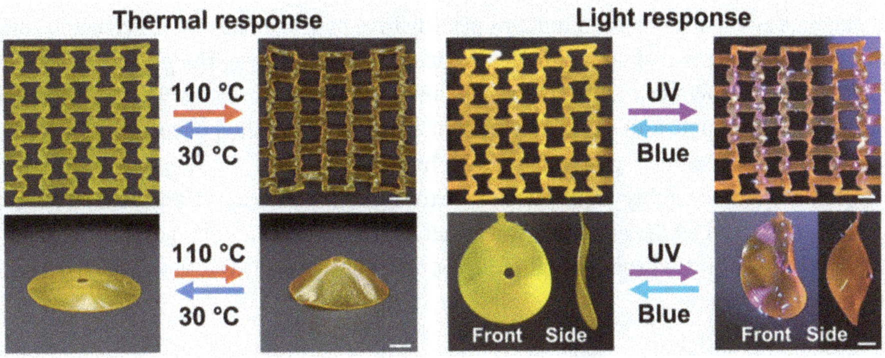

Figure 5.8: From Lugger et al.: Reversible actuation of the complex reentrant honeycomb (top) and spiral director (bottom) structures obtained by DIW in response to temperature (left) and light (right). The reentrant honeycomb structure is fixed to a rigid frame (left and right sides of the structure in the photographs). The scale bars represent 0.25 cm. Reproduced with permission.

Figure 5.9 shows a summary of thermally responsive mesogens as a function of the activation temperature in degrees Celsius (Jiang et al., "Liquid crystal elastomers for actuation: A perspective on structure-property-function relation"). Additionally, several groups are presented, which respond to specific activation wavelengths of light, which typically induce a change in *trans-* or *cis*-configuration of azo groups. Additionally, an

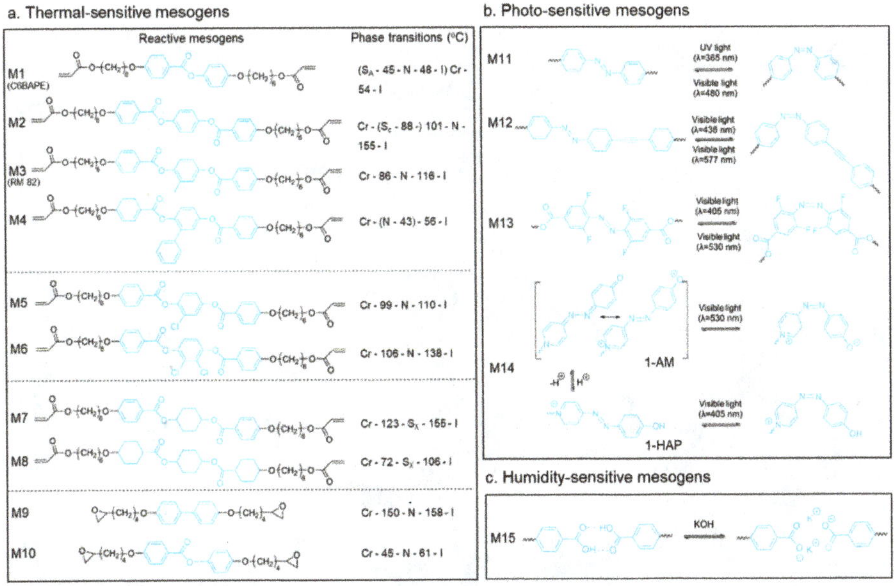

Figure 5.9: From Jiang et al.: Summary of thermally responsive mesogens as a function of actuation temperature and optic and humidity responsive systems. The activation wavelength is given for the optically driven systems.

example is given of a mesogen that has pH-sensitive hydrogen bonds, which can be disrupted by the addition of potassium hydroxide or any other base. The pH change causes formation of a polymeric salt that breaks the order of the mesogens. The base-treated LCEs are hygroscopic and exhibit significant anisotropic swelling perpendicular to the alignment direction in response to ambient humidity.

The amount of energy required to cause the transition is proportional to the amount of work produced by the actuator. As a result, thermal and humidity responsive actuators have the ability to produce higher forces compared to optical-responsive structures at equivalent strains.

5.7 Example laboratory session

In this laboratory session, students will test a soft thermally responsive material, volatile liquid-infilled elastomers. The goal of the lab is to give them an understanding of how energy is converted from a heat source into a phase change that triggers robotically relevant movement.

The materials required to make and test thermal actuators are listed in Figure 5.10 and include:

- elastomer part A,
- elastomer part B,
- ethanol,

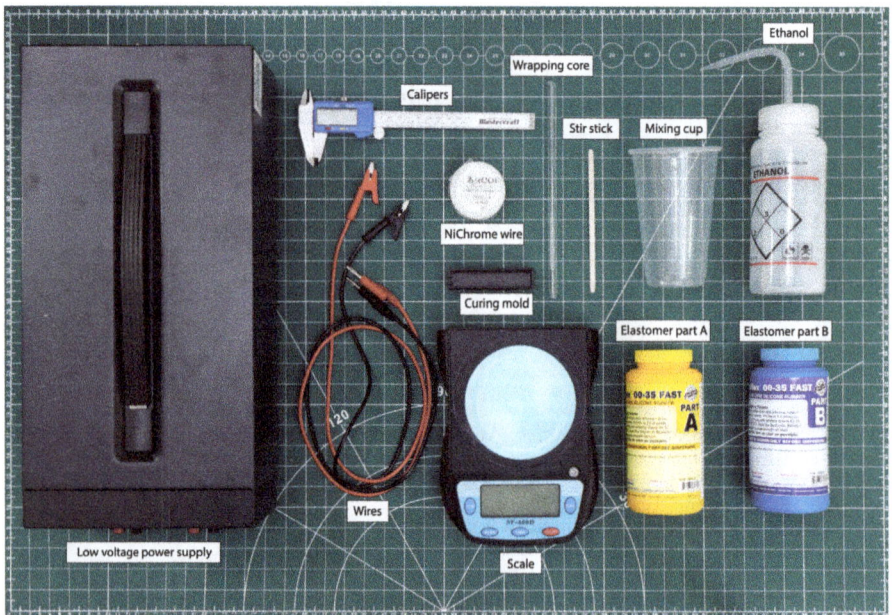

Figure 5.10: Required components for the assembly of thermally responsive ethanol-infused elastomers.

- mixing cup with a stir stick,
- low-voltage high-current power supply,
- scale,
- rulers and calipers to track displacement,
- NiChrome wire with mandrel for coiling, and
- 3D printed rectangular mold.

To measure mechanical strain in response to thermal energy, we need to be able to deliver heat in a predictable fashion. For this experiment, we will use two anisotropic methods (e. g., a hot plate and a heat gun) and an isotropic one (with a power supply, limited in uniformity by how well the coils are wrapped).

The steps to follow in the production of the elastomers are shown in Figures 5.11 and 5.12:

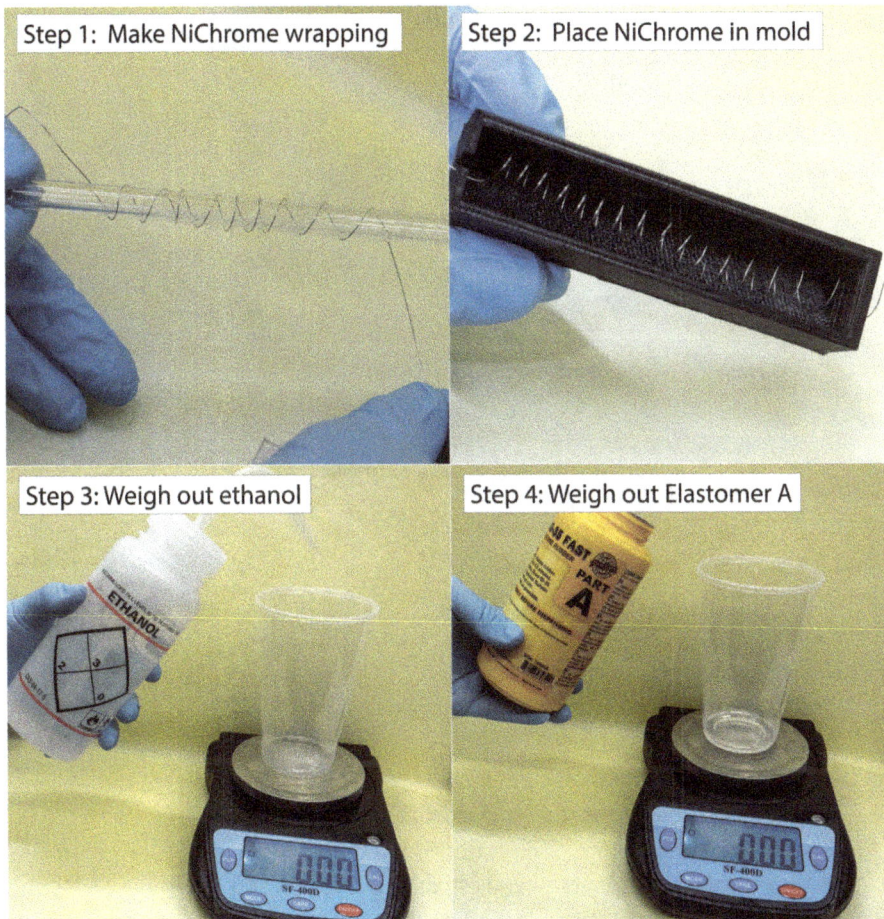

Step 1: Make NiChrome wrapping

Step 2: Place NiChrome in mold

Step 3: Weigh out ethanol

Step 4: Weigh out Elastomer A

Figure 5.11: Steps 1–4 in production of thermally responsive actuators.

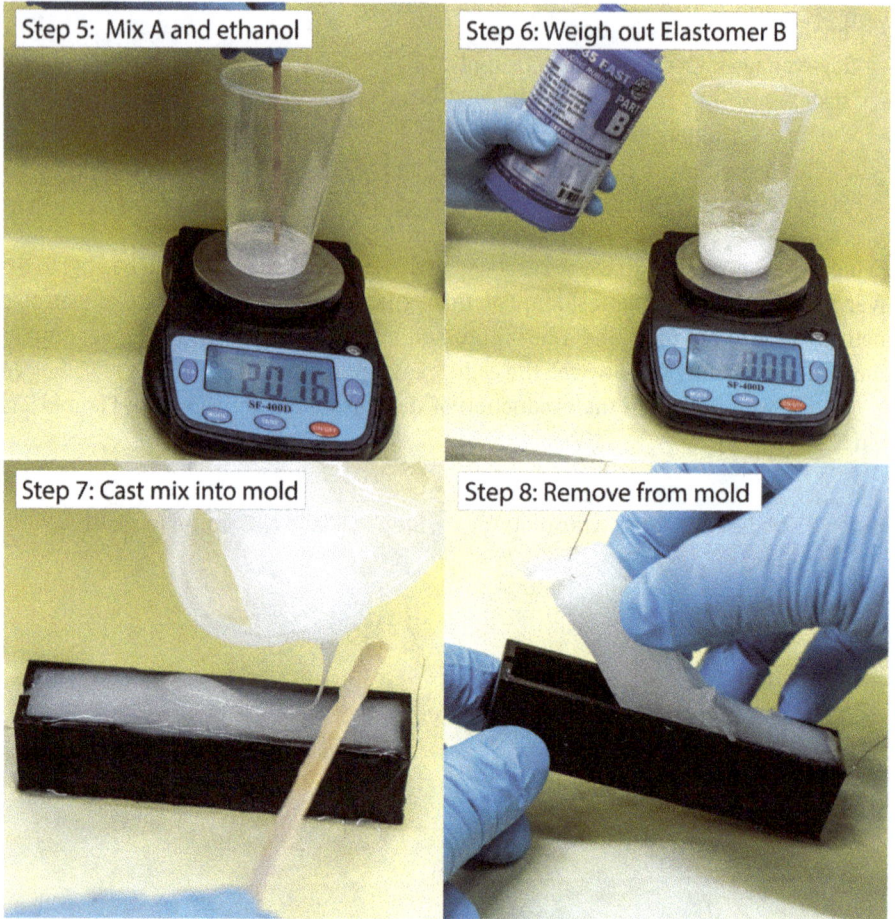

Figure 5.12: Steps 5–8 in production of thermally responsive actuators.

1. Wrap NiChrome wire around mandrel.
2. Place NiChrome coil in 3D printed mold.
3. Weigh out ethanol (20 grams).
4. Measure elastomer A (40 grams).
5. Mix ethanol and elastomer A with a stir stick.
6. Measure elastomer B (40 grams).
7. Mix A, B, and ethanol and cast into mold.
8. Cure at room temperature and remove from mold.

Figure 5.13 shows the actuator at rest and activated by the power supply (under a 9-V and 1.5-A signal). The actuators expand in all directions but may bend if one side is thicker due to a variation in the amount dispensed at the top in the curing process. One solution

Figure 5.13: Actuation of a thermally responsive actuator.

is to use the braiding approach from the McKibben actuator laboratory example and cause the actuator to contract along its long axis and operate as an artificial muscle.

The students should use the following equation for their calculations:

$$t = \frac{m \times C \times \Delta T}{\eta \times V \times I},$$
(5.12)

where m is the mass of the actuator, C is the specific heat of the Ecoflex (assume it to be $C = 1558\,\mathrm{J\,kg^{-1}\,K^{-1}}$), ΔT is the change in temperature from ambient to evaporation (assume it to be 60 K), η is the thermal efficiency (assume it to be 100 %), and V and I are the voltage and current parameters used. As part of the laboratory study, using the equation above, students should estimate the time to maximum deformation from the measured mass, voltage, current, and other given parameters. Compare the predicted time to the measured one from the table above. Give some reasoning for the difference in times.

5.8 Example problems

Problem 1. Janus fibers are comprised of two dissimilar materials joined along an interface and can be made into thermally active materials. Figure 5.14 shows an example cross-section for a Janus fiber made from materials A and B. For this problem, the active material is a shape memory polymer (SMP), and the passive material is PET.

1.   If the entire fiber diameter is 200 microns, what is the longest distance heat has to travel to activate the entire SMP?
2.   Assume that the heat diffusion coefficient between the two materials is similar $(2 \times 10^{-7}\,\mathrm{m^2/s})$. How long does it take to heat up the fiber from a uniform source?

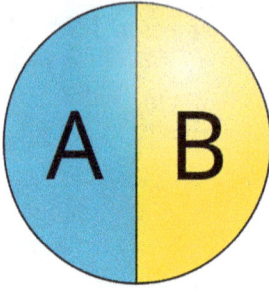

Figure 5.14: Cross-section of a Janus fiber.

3. The fiber is made by prestretching the PET and adhering it to the SMP. The SMP Young's modulus is 2 MPa, the PET modulus is 1 MPa, and the two half fibers have the same dimensions. How much prestrain should be applied to the PET such that at rest the SMP is experiencing a compressive prestrain of 5 %?

Solution. Part 1: Determine the longest distance for heat to travel.

The fiber consists of two joined materials: shape memory polymer (SMP) and PET with a total diameter of 200 µm. Since the two materials are joined along their length, the longest distance heat has to travel to activate the entire SMP is from the interface to the outer edge of the SMP. Since the fiber is split into two equal halves, the SMP occupies half of the diameter:

$$\frac{200 \ \mu m}{2} = 100 \ \mu m = 1.0 \times 10^{-4} \ m.$$

Thus the longest heat diffusion distance is 100 µm or 1.0×10^{-4} m.

Part 2: Calculate the time required for heat diffusion.

Heat diffusion in a material follows the relation

$$t = \frac{L^2}{4\alpha},$$

where $L = 1.0 \times 10^{-4}$ m (the longest distance for heat travel), $\alpha = 2 \times 10^{-7}$ m^2/s (thermal diffusivity of SMP and PET), and the factor 4 in the denominator accounts for diffusion from all sides.

Substituting the values

$$t = \frac{(1.0 \times 10^{-4})^2}{4 \times (2 \times 10^{-7})},$$

$$t = \frac{1.0 \times 10^{-8}}{8.0 \times 10^{-7}},$$

$$t = 0.0125 \ s.$$

Thus the time required for heat to activate the SMP is 12.5 milliseconds, a great response time for a thermal actuator.

Part 3: Determine the prestrain applied to PET.

We need to determine the prestrain applied to PET such that the SMP experiences a compressive prestrain of 5 % when the system is at rest. Given that Young's modulus of SMP, E_{SMP} = 2 MPa, and Young's modulus of PET, E_{PET} = 1 MPa, the SMP should experience a 5 % compressive strain (ϵ_{SMP} = −0.05). The two materials share the same cross-section and are perfectly bonded (so they experience equal and opposite forces).

Since the two materials are bonded together, the force balance equation is

$$E_{SMP} \cdot A \cdot \epsilon_{SMP} + E_{PET} \cdot A \cdot \epsilon_{PET, \, final} = 0.$$

Canceling the area A,

$$E_{SMP} \cdot \epsilon_{SMP} + E_{PET} \cdot \epsilon_{PET, \, final} = 0.$$

Substituting values,

$$(2 \times -0.05) + (1 \times \epsilon_{PET, \, final}) = 0,$$
$$-0.10 + \epsilon_{PET, \, final} = 0,$$
$$\epsilon_{PET, \, final} = 0.10 \ (10 \ \%).$$

Thus a 10 % prestrain should be applied to PET before bonding to ensure that the SMP experiences a 5 % compressive strain at rest.

Problem 2. One challenge in operating shape memory materials is to ensure the material returns to a pre-strained state after it cools down at the end of an actuation cycle. One approach are coaxial fibers, comprised of one active material, a shape memory polymer (SMP) shell, coated onto a passive material is PET (fiber core).
1. If the entire fiber diameter is 200 microns, and the SMP coating is 50 microns thick, what is the longest distance heat has to travel to activate the entire SMP?
2. Assume the heat diffusion coefficient between the two materials is similar ($2 \times 10 \ m^2/s$), how long does it take to heat up the fiber from an external uniform source?
3. The fiber is made by pre-stretching the PET, and coating on the SMP. The SMP Young's modulus is 1 MPa, the PET modulus is 2 MPa. How much pre-strain should be applied to the PET such that at rest the SMP has a compressive strain of 3 %?

Problem 3. An actuator containing pockets of ethanol dispersed in Eco-Flex 00-50 is heated up resistively. Estimate the strain of the elastomer, if the volume fraction of ethanol is 20 %, dispersed in uniform spherical pockets, and the modulus of the elastomer is 500 kPa. Assume the elastomer is heated above the boiling point of the ethanol, which expands to 9.5× the original size of the liquid.

Bibliography

Zhi-Chao Jiang et al. "Liquid crystal elastomers for actuation: A perspective on structure-property-function relation". In: *Progress in Polymer Science* (2024), p. 101829.

Sean JD Lugger et al. "4D printing of supramolecular liquid crystal elastomer actuators fueled by light". In: *Advanced Materials Technologies* 8.5 (2023), p. 2201472.

Aslan Miriyev, Kenneth Stack, and Hod Lipson. "Soft material for soft actuators". In: *Nature communications* 8.1 (2017), p. 596.

6 Magnetic soft machines

Contents

Magnetic actuators have been widely studied due to the availability of magnetically responsive materials that can generate deformation or force under the influence of magnetic fields. This chapter focuses on two key mechanisms of magnetic actuation: magnetorheological effects in fluids embedded in elastomers and reconfigurations of magnetic particle networks in soft composites, such as magnetoactive elastomers.

6.1 Learning objectives

The main concept to be presented and understood is how magnetic actuation requires an input of energy proportional to the material's volume, magnetic susceptibility, and other composite properties. Two classes of soft magnetic actuators are discussed: magnetorheological fluids embedded in elastomeric matrices (which stiffen or flow in response to a magnetic field) and soft composites with embedded magnetic particles (which produce complex deformations through controlled reconfigurations). For all these systems, the two key challenges are precise fabrication and delivery of magnetic energy to achieve effective actuation.

6.2 Background and principles

The concept of magnetic actuation in soft matrices has been studied for decades, with the earliest inventions coming from the automotive field. The first demonstrations used magnetorheological (MR) elastomers, where magnetizable components have been embedded in the elastomer during the cross-linking process, with the ultimate goal of producing automotive parts that did not require seals. Many other effects (e. g., magnetoresistivity, piezoresistive magnetodielectric, deformation, striction, magnetooptic, and magnetoacoustic) have been studied since, and such materials are now referred to with the expanded term of magnetoactive elastomers (MAEs). More recently, the development of soft magnetoactive elastomers, pioneered by researchers exploring soft

https://doi.org/10.1515/9783111069418-006

robotics, has enabled the creation of actuators capable of programmable, flexible, and complex deformations under controlled magnetic fields, opening new possibilities for soft machines and adaptive systems.

A *magnet* is a material or object that produces a magnetic field. This magnetic field is invisible but is responsible for the most notable property of a magnet: a force that pulls on other ferromagnetic materials, such as iron, steel, nickel, cobalt, etc., and attracts or repels other magnets. *Magnetization* is the vector field that expresses the density of permanent or induced magnetic dipole moments in a magnetic material. We define magnetization as the quantity of magnetic moment per unit volume, represented by a vector $\mathbf{M}$. Figure 6.1 illustrates the difference between soft and hard magnetic materials, as a function of their response to an applied magnetic field. There are two relevant magnetic fields, $\mathbf{B}$, also known as the magnetic flux density, and the magnetizing field $\mathbf{H}$. In the macroscopic formulation of electromagnetism, $\mathbf{H}$ is generated around electric currents and displacement currents and also emanates from the poles of magnets. The units of $\mathbf{H}$ are amperes per meter. The two magnetic fields help define *permeability* μ as the measure of magnetization produced in a material in response to an applied magnetic field:

$$\mathbf{B} = \mu\mathbf{H}. \tag{6.1}$$

Mathematically, the relationship between magnetization and the magnetic fields is expressed using μ_0, the magnetic permeability of vacuum:

$$\mathbf{H} = \frac{1}{\mu_0}\mathbf{B} - \mathbf{M}. \tag{6.2}$$

Magnetic susceptibility χ is a dimensionless proportionality constant that indicates the degree of magnetization of a material in response to an applied magnetic field. The

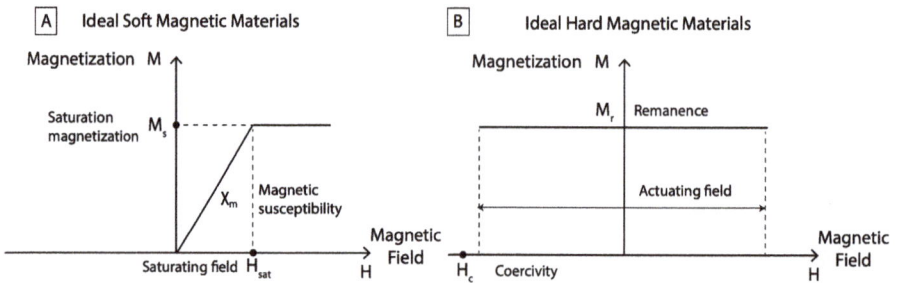

Figure 6.1: Idealized magnetic constitutive laws for soft-magnetic and hard-magnetic materials. (A) Ideal soft-magnetic materials are characterized by the linear relationship between the induced magnetization and the magnetic field with constant magnetic susceptibility before saturation and constant magnetization after saturation without magnetic hysteresis (zero remanence and coercivity). (B) Ideal hard-magnetic soft materials are characterized by large magnetic hysteresis (high remanence and coercivity) to maintain the remanence under an actuating field below the coercivity.

proportionality relation is written as

$$\mathbf{M} = \chi\mathbf{H}, \tag{6.3}$$

where **M** is the magnetization of the material (the magnetic dipole moment per unit volume) with unit amperes per meter, and **H** is the magnetic field strength, also with the unit amperes per meter.

Coercivity, also called the magnetic coercivity, coercive field, or coercive force, is a measure of the ability of a ferromagnetic material to withstand an external magnetic field without becoming demagnetized. The units for coercivity are oersted or ampere/meter units, and the quantity is denoted $\mathbf{H}_C$. *Remanence* or remanent magnetization or residual magnetism is the magnetization left behind in a ferromagnetic material after an external magnetic field is removed. Figure 6.2 illustrates the difference between soft magnetic, hard magnetic and superparamagnetic materials in terms of their coercivity and hysteresis in response to an applied magnetic field.

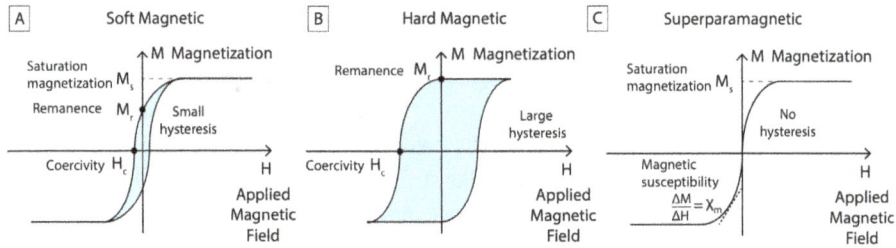

Figure 6.2: Magnetic components of magnetic soft materials can be divided into three categories (soft-magnetic, hard-magnetic, and superparamagnetic) depending on their magnetization characteristics. In general, soft-magnetic materials are characterized by their high saturation magnetization M_s, low coercivity H_c, and low remanence (M_r) with narrow hysteresis curves, whereas hard-magnetic materials are characterized by large hysteresis due to their high coercivity and remanence. Superparamagnetic materials exhibit no hysteresis and become quickly saturated under relatively low fields.

6.3 Magnethorheological fluids in soft matrices

The operation mechanism of soft materials incorporating magnetorheological (MR) fluids in elastomers is based on the tunable rheological properties of MR fluids in response to an external magnetic field. These materials, often referred to as magnetorheological elastomers (MREs) or MR fluid-elastomer composites, combine the elastic properties of the polymer matrix with the field-responsive behavior of the MR fluid. Each component serves a different function in the composite: the elastomer provides elasticity, allowing the material to spring back to its original shape after a stress is applied, and the magnetorheological fluid consists of soft-magnetic micron-sized particles (such as iron or iron-based alloys) dispersed in a carrier fluid (e. g., silicone oil). Occasionally, stabilizers and surfactants may be included to prevent particle settling and aggregation. Figure 6.3 captures the way in which MREs resist deformation, such as shear, when a stress is applied in conjunction with an applied electric field.

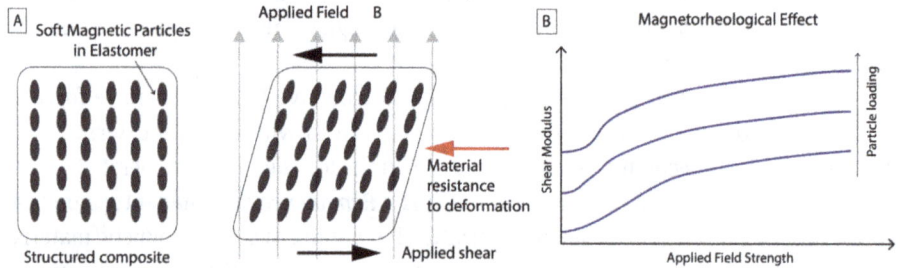

Figure 6.3: Field-induced stiffening (magnetorheological effect) of anisotropic soft-magnetic soft materials with chained particles. (A) Particles cause the composite to resist deformation under shear when a perpendicular field is applied. (B) Shear modulus increases both with higher loading of magnetic particles and higher applied fields.

When no magnetic field is applied, the MR fluid particles remain randomly dispersed within the elastomer matrix, allowing the material to behave like a soft, flexible rubber. Upon application of a magnetic field, the iron particles inside the MR fluid align along the field lines, forming chain-like structures. This dramatically increases the stiffness, damping, and shear modulus of the composite material. The result for MREs is tunability of material properties in reaction to applied magnetic fields and anisotropy. The strength of the applied magnetic field determines the extent of particle alignment, allowing for real-time control of the stiffness and viscoelastic behavior. The anisotropic mechanical behavior causes properties to be different in directions parallel and perpendicular to the applied magnetic field.

From an application perspective, the response time is typically in the millisecond range, making MREs suitable for rapid, dynamic environments. When the magnetic field is removed, the particle structures break down due to Brownian motion and elastic forces from the matrix, returning the material to its original soft state. This reversible transition and return to original shape allow for repeated tuning of the material properties required for applications.

6.4 Magnetoactive elastomers

A different type of magnetic actuator uses *magnetostriction*, the phenomenon of volume change in response to an applied magnetic field due to interactions between the embedded magnetic particles and the elastomeric matrix. This phenomenon allows for tunable mechanical properties, shape changes, and adaptive stiffness in the material.

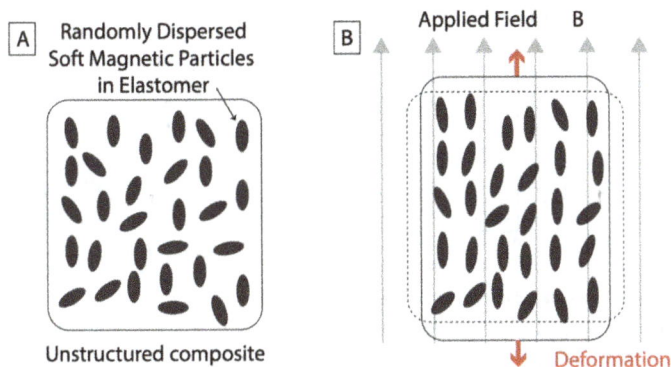

Figure 6.4: Field-induced deformation (magnetostriction) of isotropic soft-magnetic soft materials with randomly dispersed particles.

The mechanism of magnetostriction in MAEs, shown in Figure 6.4, as a contraction example, is straightforward. During the initial state, when no magnetic field is applied, the magnetic particles are randomly distributed within the soft elastomer matrix. These particles are most often metals, such as iron, cobalt, or nickel, or oxides such as iron oxide. The composite material behaves like a doped elastomer with properties different from those of a pure rubber because of blending with rigid particles. At rest, the composite shape and mechanical properties are determined solely by the polymer and particle structure and arrangement. When a magnetic field is applied, it induces magnetic dipole interactions between the particles. These interactions lead to particle realignment (see Figure 6.4 B), resulting in changes in the composite, at both the micro- and macroscales. Multiple deformation modes are possible:

- *Positive magnetostriction:* when particles align along the magnetic field lines, the material can elongate in the field direction.
- *Negative magnetostriction:* if the elastomer matrix restricts the particle movement, then internal stresses can cause contraction instead of expansion.
- *Shear deformation:* in anisotropic MAEs, where the particles are aligned in a preferred direction, applying a transverse magnetic field can cause bending or twisting instead of pure elongation or contraction.

The behavior of the composite as a function of magnetic particle loading is described in detail in Figure 6.5. The actuator performance has two set boundaries, which have a clear trade-off between themselves: the magnetization and the Young's modulus of the composite. As shown in Figure 6.5 A, the magnetization of the composite (M) increases monotonically with increased volume fraction of magnetic particles. This response is not surprising, as, on a per volume basis, the composite contains more and more magnetically responsive particles. However, the stiffness of the magnetic particles, which is higher than that of the elastomeric matrix, causes the Young's modulus of the composite (Y) to increase exponentially, as shown in Figure 6.5 B.

Composite Material Properties due to Particle Loading

Actuator Performance Variation due to Particle Loading

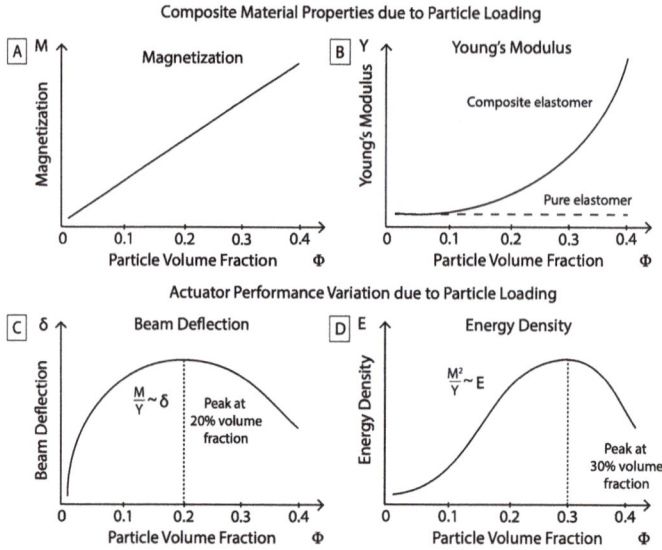

Figure 6.5: Design optimization of magnetic composites based on soft polymers with embedded hard-magnetic particles. (A–B) Material properties (magnetization and shear modulus) of magnetic soft composites varying with the particle volume fraction. (C–D) Actuation performance of magnetic soft bending actuators in terms of the free-end deflection and energy density varying with the particle volume fraction.

The stiffness at higher-volume fraction of magnetic particles causes the relevant actuation properties to have inherent maximums. For applications where large deformation is needed, the composite material should be as soft as possible. As an approximation, Figure 6.5 C indicates that a composite beam will show the largest deformation δ when the particle loading is approximately 20 %. This maximum is due to the interdependence of deflection to the material stiffness and magnetization:

$$\delta \approx \frac{M}{Y}. \tag{6.4}$$

For applications where energy density needs to be maximized is needed, the composite material should contain an appropriate ratio of magnetic particles while not being too stiff. As an approximation, Figure 6.5 D indicates that a composite actuator will show the largest energy density E when the particle loading is approximately 30 %. This maximum is due to the different dependence of energy density to the material stiffness and magnetization:

$$E \approx \frac{M^2}{Y}. \tag{6.5}$$

Under these constraints, there is an additional control parameter over the actuation modality, the type of magnetic field applied. Applying a uniform magnetic field on a magnetoactive elastomer causes a torque on the magnetic particles and aligns them with the direction of the field. In contrast, applying a nonuniform magnetic field causes both a torque and a force to be applied to the individual particles. Figure 6.6 (Kim and Zhao, "Magnetic soft materials and robots") shows the difference at the particle vs. macroscale level using the example of a bending beam. As shown in **b**, a rectangular beam made of a hard-magnetic composite with a deformable elastomer experiences torque driven actuation. The beam is uniformly magnetized along the length direction and deforms under a uniform actuating field that is applied perpendicularly to the beam's remanent magnetization. The schematic captures that applying a nonuniform magnetic field, as exemplified in **d**, causes relatively more bending in the beam due to the resultant force. The bending actuation of the composite beam under nonuniform actuating fields has a similar initial response due to the produced torque. However, the actuation is increased by the magnetic force as the body deforms to align its remanent magnetization with the applied field.

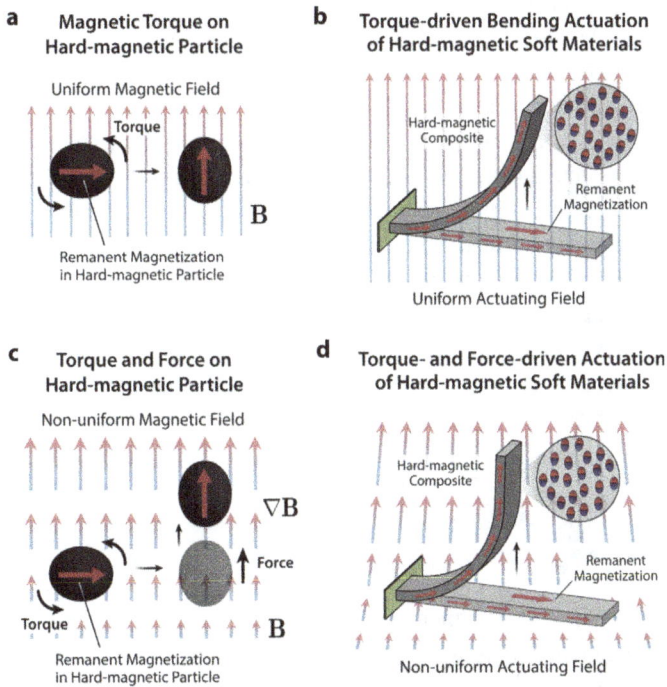

Figure 6.6: From reference, (a) magnetic torque acting on a magnetized hard-magnetic particle under a spatially uniform magnetic field. (b) Response at the macroscale in a bending beam under torque. (c) Magnetic torque and force acting on a magnetized hard-magnetic particle under a spatially nonuniform magnetic field, in which the particle not only rotates due to the magnetic torque but also moves toward the direction of the increasing field due to the attractive magnetic force. (d) Response at the macroscale in a bending beam under torque and force.

With these building blocks, magnetoactive elastomers can be arranged and programmed in a broad range of approaches, well beyond the scope of this book. Additional considerations for reliable operation of these devices are the following:

–   The interface between magnetic particles and elastomers may be prone to *corrosion* in wet environments. Approaches to mitigate this include primarily coatings of silica, which can be bonded to the magnetic particles in a variety of ways to improve stability.

–   Magnetic materials can be *demagnetized* when heated above their Curie temperature, which allows for reprogramming of magnetic moments under heat and applied fields. This approach is useful in creating complex patterns that drive complex deformation modes.

–   The magnetoactive elastomers respond to applied fields, which can be produced either by magnets manipulated by *multidegree-of-freedom machines*, or by manipulation platforms based on stationary *multiaxial electromagnets* in nonorthogonal eight-coil configurations and commercially available systems. Both approaches have logistical challenges and are actively explored in research.

–   The ability to actuate magnetic composites with external fields allows for fantastic versatility in *untethered robotics* (showcased in Figure 6.7), as well as **programmable surfaces, active materials, and origami**. These types of devices have been shown to be useful in active transport of droplets, locomotion in unstructured environments, tensegrity and auxetic structures, and tunable surface properties.

–   Hybrid approaches can be accomplished by combining magnetic materials with shape memory materials, as described in Chapter 5, flexible electronics from Chapter 2, and swellable and deswellable hydrogels.

–   Applications of magnetic soft machines are primarily in the medical field, where they have been explored as magnetic hydrogels for on-demand drug or cell delivery based on field-induced deformation, magnetic soft capsule robots with embedded magnets for targeted drug delivery and endoscopic imaging, and magnetically steerable end effectors for imaging and intervention.

6.5 Alternative uses: magnetic sensing and adhesion

Due to their structure and properties, magnetoactive elastomers can also be designed to serve as sensors and controllable adhesion systems. From a sensing perspective, the MAEs can operate as strain or deformation sensors. When stretched, compressed, or bent, the alignment of magnetic particles inside the elastomer changes, altering the overall magnetic properties. The change can be detected using magnetic field sensors (e. g., Hall effect sensors) in an adjacent structure and have applications in touch sensors, wearable motion sensors, etc. Similarly, the system can be adapted to serve as a pressure or force sensor, as the force changes the spacing of magnetic particles, leading to a measurable shift in the internal magnetic field. Additional applications include

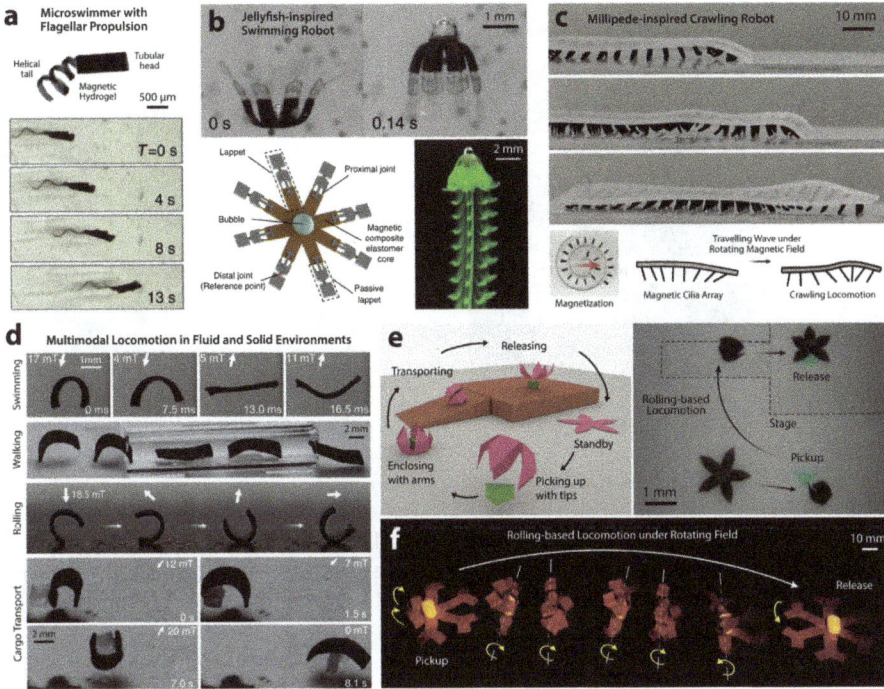

Figure 6.7: From Kim et al., small-scale untethered soft robots based on magnetic soft materials. (a) Microswimmer based on magnetic hydrogel mimicking the helical propulsion of bacterial flagella under rotating magnetic fields. (b) Bioinspired magnetic soft robots based on hard-magnetic soft composites mimicking the swimming motion of a jellyfish under alternating magnetic fields. (c) Millipede-inspired crawling robot with an array of hard-magnetic cilia with different magnetization directions to produce traveling waves under a rotating magnetic field for crawling locomotion. (d–f) Magnetic soft robots based on hard-magnetic composites exhibiting multimodal locomotion such as swimming, walking, and rolling in fluid or solid environments as well as cargo transport through spatiotemporal control of the actuating magnetic fields.

structural health monitoring in smart materials. In reverse order, the devices can be used to detect magnetic fields by measuring the accompanying change in shape due to the external fields. Additional hardware is required to detect the shape change, but new applications are possible, including proximity sensors for industrial automation, navigation aids for magnetic field mapping, and security and antitampering devices.

Additionally, MAEs can be used to create controllable adhesion surfaces by modulating their stiffness, surface roughness, or interaction with magnetic fields. When considering direct magnetic adhesion, the MAEs can stiffen in response to an applied magnetic field, increasing their contact area and adhesion to a surface. When the field is removed, the material softens, reducing adhesion and enabling easy detachment. However, more interesting behavior can be obtained when magnetic materials are combined with tailored adhesives, such as gecko-inspired deformable microstructures. Figure 6.8

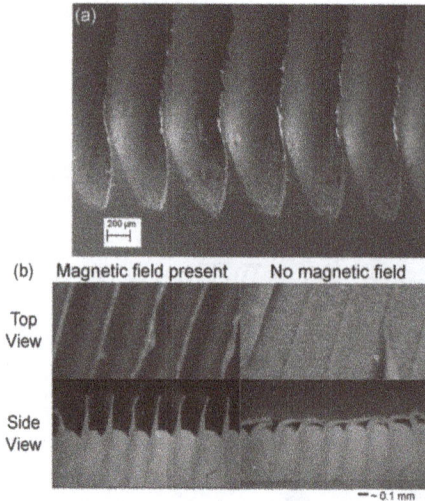

Figure 6.8: From Gilles et al., SEM of the completed magnetoelastomer microridges. The microridges are 325-μm long, 15-mm wide, and taper from 100 μm at their base to less than 10 μm at the tip. (b) Actuation of the ridges, as seen from the top and side. In the presence of a magnetic field, the ridges completely flatten. Reproduced with permission.

shows an early example (Gillies, Kwak, and Fearing, "Controllable particle adhesion with a magnetically actuated synthetic gecko adhesive") of a magnetic material made into microridges. When a field is applied, the ridges flatten to increase the contact area between the soft material and its surroundings. The approach demonstrates controllable adhesion to glass spheres with a new magnetically actuated synthetic gecko adhesive made from a magnetoelastomer composite. Capable of controlling adhesion to glass spheres 500 μm to 1 mm, this represents an important step in realizing an adhesive with dry self-cleaning capabilities across a wide range of particle sizes. Chapter 8 will discuss gecko-inspired microstructures in more detail.

6.6 Example laboratory session

In this laboratory session, students will test a soft magnetically responsive material, iron particle infilled elastomers. The goal of the lab is to give them an understanding of the ability of the material to respond to an applied magnetic field directly influenced by its composition.

The materials required to make and test magnetic actuators are listed in Figure 6.9 and below and will include:
– elastomer part A,
– elastomer part B,
– iron fillings,

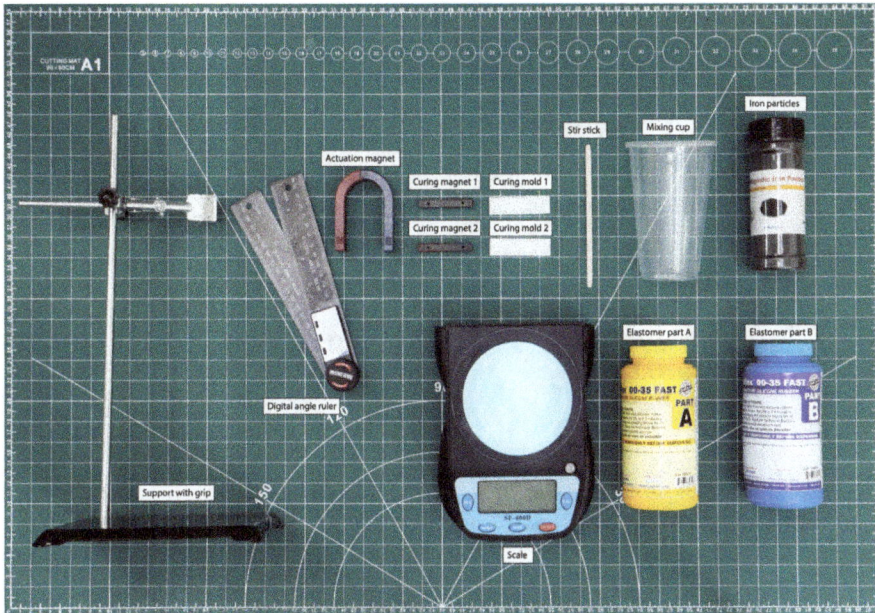

Figure 6.9: Required components for the assembly of magnetically responsive soft actuators.

- mixing cup with a stir stick,
- set of magnets for particle alignment and actuation,
- scale,
- digital angle ruler to track bending,
- 3D printed rectangular molds, and
- laboratory stand with grip.

The steps to follow in the production of the magnetically responsive elastomers are shown in Figures 6.10 and 6.11. One suggestion is to have different groups make composites at different volume loadings of iron particles to aim to reproduce the behavior predicted by the graph in Figure 6.5 C.

1. Weigh out iron fillings (amount to vary by student group).
2. Measure elastomer A.
3. Mix iron fillings and elastomer A with a stir stick.
4. Measure elastomer B.
5. Mix A, B, and iron and cast into mold, with or without magnets nearby to cause particle alignment.
6. Cure at room temperature and remove from mold.
7. Secure the end of the bending beam with the lab support.
8. Measure beam deflection in response to actuation magnet brought in vicinity of the beam.

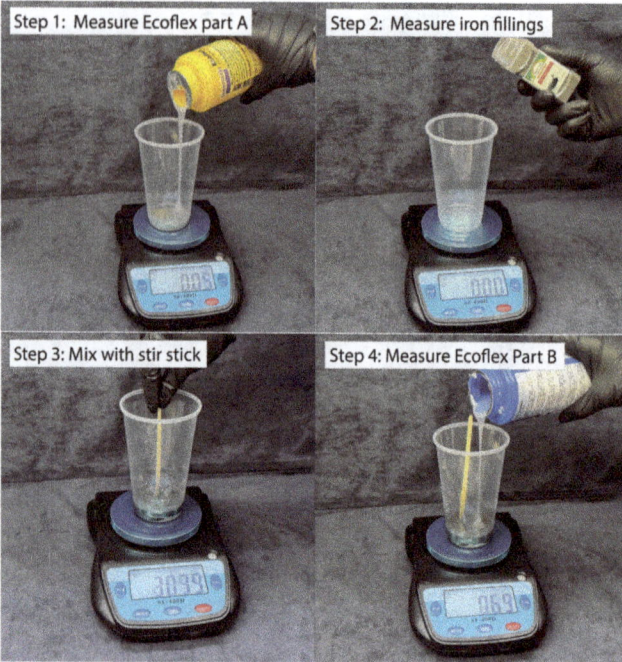

Figure 6.10: Steps 1–4 in production of magnetically responsive actuators.

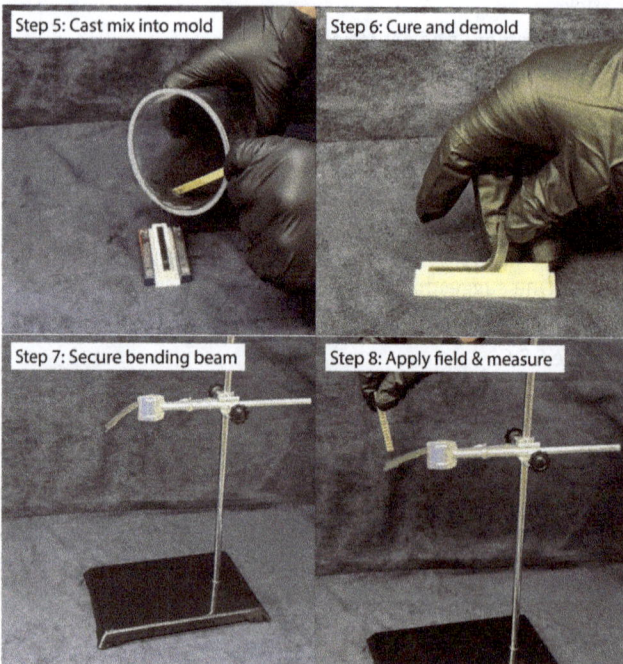

Figure 6.11: Steps 5–8 in production of magnetically responsive actuators.

Note. An alternative method of measuring magnetostriction is shown in Figure 6.12. The composite beam is secured with small magnets to one side of a pair of calipers. The calipers are slowly closed, and the beam stretches when close enough to respond to the field of magnets attached to the opposite side of the calipers. The yellow circles show the reader where to focus in measuring length change and strain.

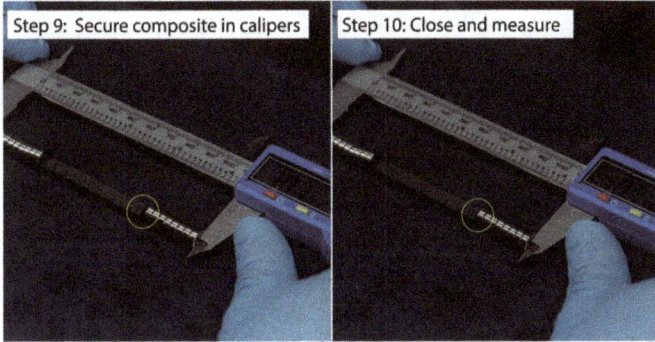

Figure 6.12: Alternative method of measuring magnetostriction.

6.7 Example problems

Problem 1. A soft elastomer composite contains 30 % by weight magnetic iron particles, uniformly dispersed within the material. The elastomer has an initial length of 10 cm in the direction of an applied magnetic field. The Young's modulus of the composite elastomer is 2 MPa, and the bulk magnetostriction coefficient of the composite is 2×10^{-4} per Tesla. A uniform magnetic field of 0.5 T is applied along the length of the elastomer. Assuming that the magnetostriction effect follows a linear relationship with the applied field, determine:
1. The change in length of the elastomer due to magnetostriction.
2. The resulting strain in the elastomer.

Solution. Part 1. Initial length of the elastomer $L_0 = 10\,\text{cm} = 0.10\,\text{m}$. Magnetostriction coefficient $\lambda_m = 2 \times 10^{-4}$ per Tesla. Applied magnetic field $B = 0.5\,\text{T}$.
Compute the length change, which due to magnetostriction is given by

$$\Delta L = \lambda_m B L_0.$$

Substituting the given values, we have

$$\Delta L = \left(2 \times 10^{-4}\right) \times (0.5) \times (0.10),$$
$$\Delta L = 1.0 \times 10^{-5}\,\text{m} = 10\,\mu\text{m}.$$

Thus the elastomer elongates by 10 μm under the applied magnetic field.

Part 2. Find the strain just for this problem. To avoid confusion with the symbol used for magnetostriction coefficient, the strain is labeled ε and defined as the change in length divided by the initial length:

$$\varepsilon = \frac{\Delta L}{L_0},$$

$$\varepsilon = \frac{1.0 \times 10^{-5} \text{ m}}{0.10 \text{ m}},$$

$$\varepsilon = 1.0 \times 10^{-4}.$$

Thus the strain in the elastomer due to magnetostriction is 1.0×10^{-4} (or 0.01 %).

Problem 2. A material can became more or less adhesive depending on the position of some magnetically-controlled microridges, as shown in Figure 6.8. A difference in the adhesive properties between the vertical and flattened positions can be explained by the change in effective stiffness between the two states. Previous studies have shown that below an elastic modulus of approximately 100 kPa, known as the *Dahlquist criterion*, the material surface becomes tacky. A stiffer material, with a modulus well above the Dahlquist criterion can be made to have an effective modulus below 100 kPa through only the geometric design in its surface. In this problem we'll estimate the reverse: making something less sticky by making it significantly stiffer.

You are working with a 1 cm × 1 cm elastomeric patch embedded with a magnetorheological (MR) fluid. The base material has an initial Young's modulus of 50 kPa, and a volume fraction of fluid of 30 %. When a magnetic field B is applied, the MR fluid responds by forming chain-like microstructures, increasing the overall stiffness of the composite material. A common model for field-dependent stiffening in MR elastomers follows:

$$\Delta E \approx \eta \phi \mu_0 M^2$$

- η is a material-dependent scaling factor (determined experimentally, typically around 10^{-1} for soft elastomers);
- ϕ is the volume fraction of MR fluid;
- μ_0 is the permeability of free space ($4\pi \times 10^{-7}$ H/m);
- M is the field-dependent magnetization of the MR particles. For small applied fields we assume linear magnetic response with susceptibility, $M \approx \chi B/\mu_0$ and for this material χ is 4.

Question: for an applied field of 0.1 T, does the material become stiff enough to stop being tacky? What about at 0.5 T?

Bibliography

Andrew G Gillies, Jonghun Kwak, and Ronald S Fearing. "Controllable particle adhesion with a magnetically actuated synthetic gecko adhesive". In: *Advanced Functional Materials* 23.26 (2013), pp. 3256–3261.

Yoonho Kim and Xuanhe Zhao. "Magnetic soft materials and robots". In: *Chemical reviews* 122.5 (2022), pp. 5317–5364.

7 Controllable adhesion in soft materials

Contents

Controllable adhesion is highly desirable in soft materials, because it enables improved locomotion and manipulation. This chapter focuses on two key mechanisms of tunable adhesion, electro-adhesion with interdigitated patterns and gecko-inspired adhesives with tunable contact surface area.

7.1 Learning objectives

The main concept to be presented and understood is how dry electro-adhesion and gecko-inspired adhesion rely on surface interactions to achieve strong yet reversible attachment. Two primary approaches are discussed: electro-adhesion, which utilizes electrostatic forces generated by applying a voltage to induce attraction between surfaces, and gecko-inspired adhesion, which mimics the hierarchical micro- to nanoscale structures of gecko setae to achieve van der Waals-based adhesion. For both systems, the two key challenges are the precise fabrication of adhesion structures and the effective control of surface forces to enable strong, reliable attachment and detachment.

7.2 Background and principles

When discussing adhesion, it refers primarily at achieving attraction between different materials, as opposed to cohesion, where similar materials attract each other. The need for adhesion appears in both locomotion and manipulation applications, in particular for tasks such as climbing, perching, and delicate handling. One of the most widely used industrial approaches, chemical adhesion, is not applicable in these circumstances because it most often causes irreversible adhesion. The basic operation modes for the most widely studied reversible dry adhesives (i. e., electro-adhesives and gecko-inspired adhesives) are shown in the next two figures. Details for each are presented in the following sections.

https://doi.org/10.1515/9783111069418-007

An *electroadhesive* is a material or surface that utilizes electrostatic forces to achieve adhesion to a contacting substrate. This adhesion is generated by applying a voltage, which induces opposite charges on the adhesive surface and the substrate, creating an attractive force (see Figure 7.1). The key feature of electro-adhesive systems is their ability to adhere to a wide range of materials, including insulating and textured surfaces, without requiring chemical bonding or complex mechanical structures. *Electro-adhesive performance* is characterized by the strength and reversibility of the electrostatic attraction, allowing for tunable adhesion. We define adhesion effectiveness as the ability to generate sufficient electrostatic force while maintaining energy efficiency and ease of detachment, typically represented by adhesion strength per unit voltage and power consumption.

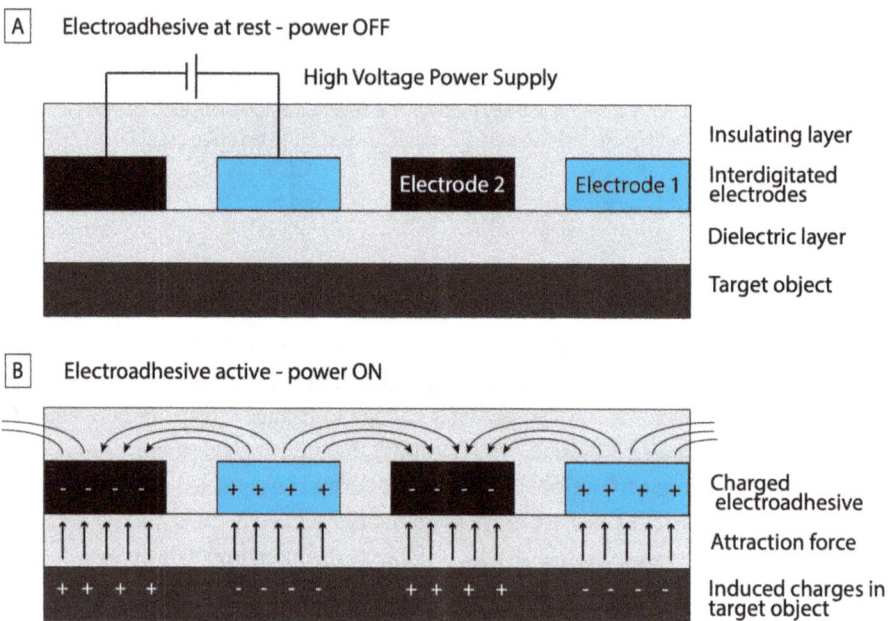

Figure 7.1: Schematic of the operation mode of an electro-adhesive based on interdigitated electrodes. *A*: cross-section of the electrode pattern, separated by the target object by a dielectric layer. **B**: when the electro-adhesive is turned on, the charges in the interdigitated electrodes induce charges in the adjacent target object. The attraction between the induced charges and those in the interdigitated electrodes cause the electro-adhesive force.

A *gecko-inspired adhesive* is a material or surface that mimics the unique adhesion mechanism of gecko feet. This adhesion is achieved through van der Waals forces, which arise from the interaction between the adhesive surface and a contacting substrate (see Figure 7.2). The key feature of these adhesives is their hierarchical micro- and nanoscale structures, often designed to maximize surface contact and enhance adhesion without

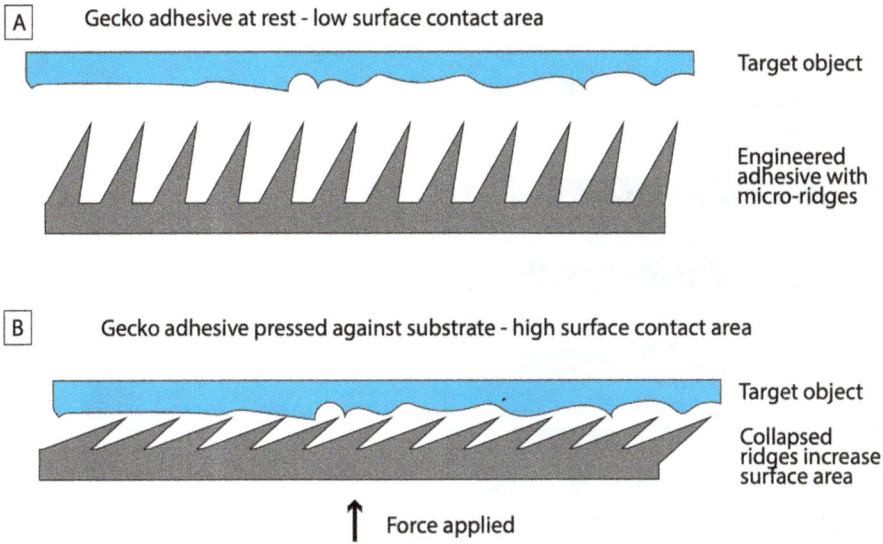

Figure 7.2: Schematic of the operation mode of a gecko-inspired adhesive based on patterned microridges. **A**: Cross-section of the microridges, separated by the target object by an air gap. **B**: When force is applied and the gecko-adhesive is brought into contact with the target object, the microridges deform to increase surface area. The increased surface area drives stronger van der Waals interactions, leading to a strong adhesive force between the objects. When the applied force is released, the microridges return to their original shape, drastically reducing the contact area and the resulting adhesion force.

requiring chemical bonding or external energy input. *Gecko-adhesive performance* is characterized by the ability of these structures to generate and control intermolecular forces, allowing for strong yet reversible attachment. We define adhesion effectiveness as the ability to maximize contact area while maintaining ease of detachment, typically represented by shear and normal force capabilities. Typically, a gecko-inspired adhesive requires an additional actuation mechanism that causes a force to be applied to the adhesive to increase contact area.

7.3 Electro-adhesion

Electro-adhesion is a dry and reversible process, which relies on electrostatic forces to create reversible adhesion between a surface and a substrate. When a voltage is applied to an electro-adhesive pad, opposite charges are induced on the substrate, generating an attractive force that holds the surfaces together. The strength of adhesion depends on factors such as applied voltage, dielectric properties of the materials, surface roughness, and electrode design. An example of an interdigitated design is shown in Figure 7.3, where some of the relevant parameters are listed, including digit length, width, spacing, and the total number of digits. Experimental studies have demonstrated that optimizing

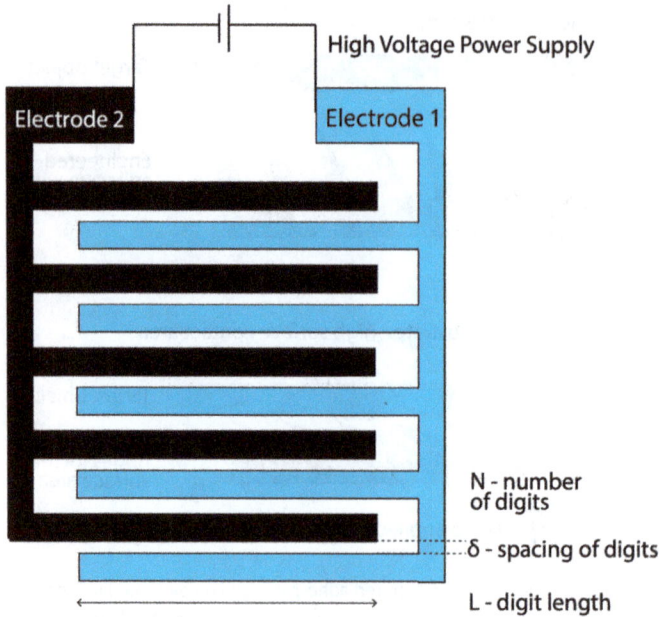

Figure 7.3: Example of the pattern required for electro-adhesion, as a top down view. The two electrodes have an interdigitated pattern, where digits from each run in parallel to the other. The length, spacing, width, and number of digits all influence the strength of the electro-adhesive force.

the electrode configuration at equivalent area can enhance the adhesion force by up to 15× compared to simple designs, with few features. The most widely adopted pattern is interdigital, which generates hemispherical fields of equal potential at the interface between electrode pairs.

In the simplest approximation, at the microscale, electro-adhesion is based on *Coulomb attraction* between charges. For continuous charge distributions and fields, the mechanism is mathematically represented by the equation

$$F = \frac{A}{2}\epsilon_0 \left(\frac{\epsilon_d V}{t} \right)^2,$$

(7.1)

where
- F is the electro-adhesive force,
- A is the contact area of the electrodes,
- t is the thickness of the dielectric,
- ϵ_0 is the permittivity of vacuum,
- ϵ_d is the dielectric constant of the dielectric material, and
- V is the applied voltage.

There is a secondary effect at work, the *Johnsen–Rahbek (JR) effect*, which occurs when charge carriers are mobile enough to cause polarization in the material. Figure 7.1 shows the behavior of a material undergoing the JR effect and becoming polarized when an electro-adhesive is powered up in its vicinity. The force due to the Johnsen–Rahbek effect has slightly different dependence on the same properties and parameters:

$$F_{JR} = \frac{A_{eff}}{2} \epsilon_0 \left(\frac{\epsilon_d V_{eff}}{t} \right)^2, \tag{7.2}$$

where
- F_{JR} is the electro-adhesive force,
- A_{eff} is the effective contact area of the electrodes and the target object,
- t_g is the thickness of interfacial gap,
- ϵ_0 is the permittivity of vacuum,
- ϵ_g is the dielectric constant of the material in the gap between the object and adhesive pad, and
- V_{eff} is the effective applied voltage, accounting for resistance in the materials and the voltage drop across the gap.

Challenges in fabrication and operation of electro-adhesives are as follows:
- *Target material compatibility:* for electro-adhesives to work, they need to be able to induce a charge polarization in the target material. As a result, they can reliably adhere to a wide range of polarizable materials, including both insulating and metallic such as glass, metal, and a range of polymers.
- *Surface roughness:* to adhere to rough surfaces, electro-adhesives need to be relatively soft, or compliant, to deform an increase contact area to the target object.
- *Electrode design:* the physics of electro-adhesion is fairly complex, and most research papers on the subject take an empirical approach to optimization, with the end goal of maximizing force.
- *Fabrication:* smaller features are desired because they typically show higher output electro-adhesive forces. Both flexible and stretchable electrode processing techniques have been applied, with a good level of success.
- *Energy and power requirements:* for effective grip, electro-adhesion requires a continuous voltage supply, typically in the 1–5 kV range. This operating range matches typical dielectric elastomer actuator power requirements, and for this reason, the two technologies are often combined, with an example shown in Figure 7.4 (Gu et al., "Soft wall-climbing robots").
- *Environmental sensitivity:* contaminant factors such as humidity, dust, and surface fouling can reduce the effectiveness of the electro-adhesive pads, making it necessary to develop coatings or adaptive control strategies to mitigate these effects.

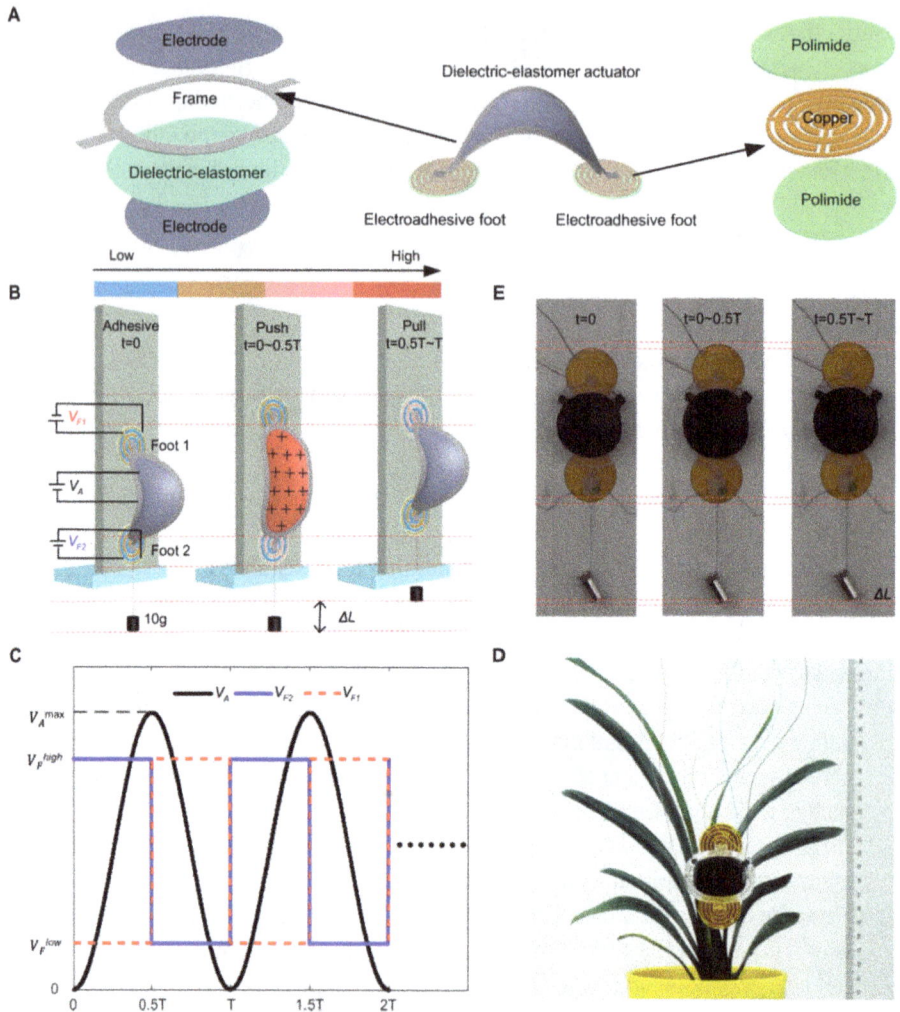

Figure 7.4: From Gu et al. 2018, example of a combined DEA-robot with electro-adhesive (EAs) pads. (A) The soft robot integrates a dielectric-elastomer actuator as its deformable body and two double-pole electro-adhesive pads as its feet, with controlled adhesion. (B) Schematic of the climbing principle. (C) Sequence of the control voltages for the actuator and electro-adhesive feet to achieve the climbing locomotion. (D) Still image of climbing upon a transparent glass wall. (E) Still images of the climbing process with a 10-g payload on the wood.

Electro-adhesives have direct applications in robotics, in particular aiding in manipulation in locomotion, as shown above, and in a broad range of environments from industrial fabrication to space manipulation. In addition, another operation mode has significant relevance for soft robotics: an *electro-adhesive clutch*, a type of clutch that utilizes electrostatic forces to control the transmission of torque between two components. By applying or removing voltage the clutch can engage or disengage, allowing for smooth, variable, and energy-efficient actuation. This variant of an electro-adhesive system enables another range of applications:

- *Prosthetic joints:* electro-adhesive clutches can provide adjustable resistance for prosthetic limbs, helping users perform smooth, natural movements without excessive energy consumption.
- *Exoskeletons for rehabilitation:* EA clutches can dynamically control resistance in assistive exoskeletons, helping users with mobility impairments by selectively locking and unlocking joints.
- *Low-power actuation in spacecraft:* electro-adhesive clutches can be used in robotic arms or deployable structures in space, where minimizing weight and power consumption is critical.

7.4 Gecko-inspired adhesives

While the gecko's adhesive system is one of the most famous, other organisms exhibit similar or complementary strategies in multiple species across a range of sizes, as shown in Figure 7.5 (Zhao, Xia, and Zhang, "A review of bioinspired dry adhesives: from achieving strong adhesion to realizing switchable adhesion"). Many insects use microscale hair-like structures (setae) on their feet to generate van der Waals forces for adhesion. Examples include leaf beetles (*Gastrophysa viridula*) and jumping spiders (*Evarcha arcuata*) using microscopic arrays of setae to cling to smooth surfaces.

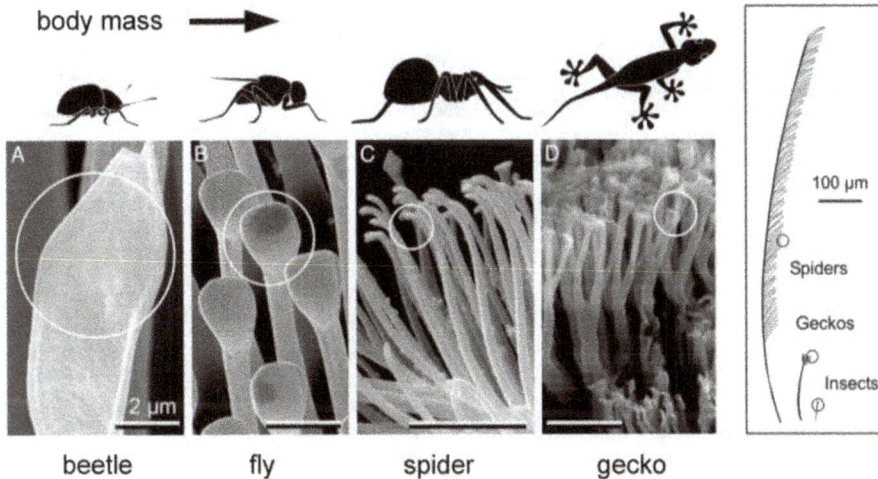

Figure 7.5: From Zhao et al., 2024. Examples of terminal elements (circles) in animals with hairy design of attachment pads. Note that heavier animals exhibit finer adhesion structures.

Reproducing natural designs, such as gecko's setae, in exact features at scale is extremely challenging from a manufacturing perspective. However, detailed analysis of the mechanism of adhesion has revealed the fundamental sequence, which needs to be replicated for dry, controllable adhesion to be achieved. The low and high adhesion states are shown in Figure 7.2 and are directly proportional to the area of adhesive in contact to the target object. When geckos want to adhere to a specific spot, they press their toes onto the surface and pull the toe pads down. The motion engages the setae and drastically increases the contact area between the pad and the target object. To release, the gecko pushes the toe back up, causing the setae to return to their original orientation and the area of contact to decrease significantly.

Robotic systems capable of dry, controllable surface-governed adhesion require several requirements to be met:

- Individual features should be in the micron range (10–500 µm), where surface forces dominate over volumetric ones.
- The microridges should be asymmetric, such that deformation in one specific direction would produce a large change in contact area.
- The structures should be made from soft materials, such as elastomers, that can both deform with ease and spring back to their original shape when the applied force is released.

Figure 7.6 shows a wide range of methods used to produce gecko-inspired dry adhesive surface based on the mechanism of shear induction: with the common theme of angled

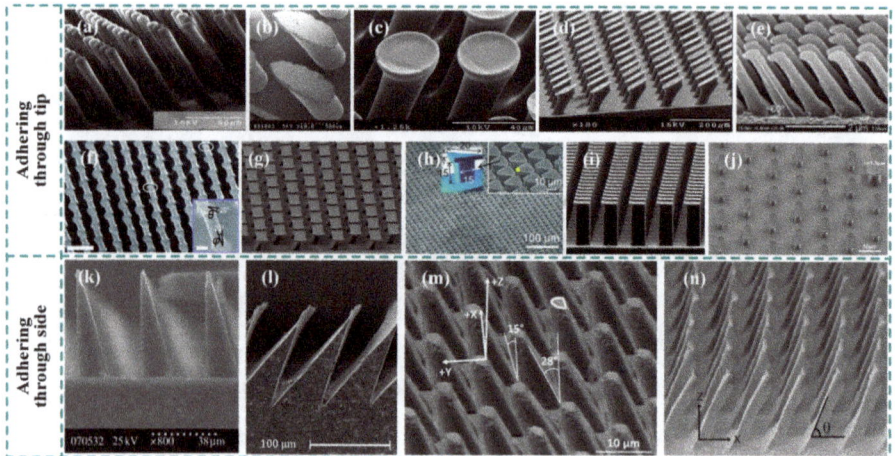

Figure 7.6: From Wang et al. 2021. Examples of artificial setae produced in a range of methods.

structures with larger features at the tip to promote van der Waals adhesion. Typically, these systems are made via lithography processes in clean room environments from polymer materials that have some flexibility under deformation. As shown at the end of Chapter 6, microridges can be fabricated via methods that require less instrumentation, such as CNC milling of wax and casting elastomer into the mold. From Wang, Liu, and Xie, "Gecko-like dry adhesive surfaces and their applications: a review".

When integrated into soft robotic systems, gecko-inspired have several advantages: primarily, they are *energy efficient*, because they do not require continuous power to maintain adhesion. The mechanism can operate in *dry and vacuum environments*, allowing for a wide range of applications. One of the main drawbacks of these adhesives is the need for a *specific activating motion*, which brings the adhesive pad in close contact with the target object and deforms the pad such that the contact area is increased drastically. Additionally, similar to living geckos, the pads struggle in wet environments and can be fouled by dust, losing adhesion power until the dust particles are removed. Hybrid approaches, such as the integration of magnetic particles in the microridges shown in Chapter 6, offers additional control over the adhesion mechanism.

7.5 Alternative approaches: octopus sucker replicas and capillarity controlled wet adhesion

One of the alternative natural approaches to control adhesion is the octopus sucker, an end effector, which can apply localized vacuum to achieve intimate contact with target objects. Engineered examples, as that shown in Figure 7.7, here combines soft adhesion actuators with supporting structures that produce a pressure change in a cavity around the target object. The soft adhesion actuator is conceptually similar to the PneuNets discussed in Chapter 3 and is composed of two layers with an embedded spiral pneumatic

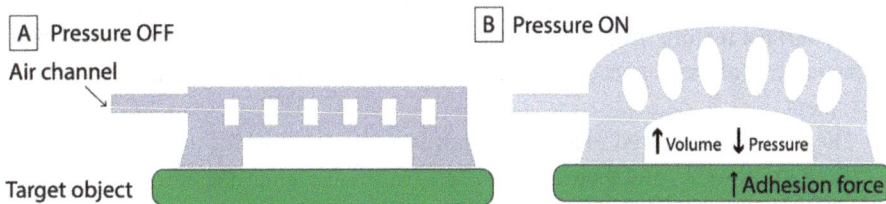

Figure 7.7: Schematic of the mechanism for switchable adhesion in the adhesion actuator upon pneumatic pressurization of a modified PneuNet. Left: adhesion-off state. Right: pressure in the chamber causes the air pocket below to change shape. The local drop in pressure causes a vacuum-driven adhesion force to the target object.

channel on top of a cylindrical chamber. This geometry allows the operator to apply positive pressure to deform the planar bilayer structured soft actuator into an inflated 3D domed shape. The deformation in the network of chambers causes deformation and a pressure drop in the cavity in contact with the target object, enabling stable and switchable adhesion.

Lastly, some organisms are able to control small liquid features on their toe pads to produce sufficient capillary force to adhere to different substrates. Specifically, tree frogs, such as *Litoria caerulea*, use mucus-secreting toe pads with a hexagonal nanopillar pattern that enhances adhesion through capillary forces due to the surface tension of the mucus. Figure 7.8 shows a simplified version of the method, where pockets of liquid deform to create a film between the toe pads and the target object. The capillarity driven adhesive system works even on wet surfaces, making it distinct from geckos, which struggle in humid conditions.

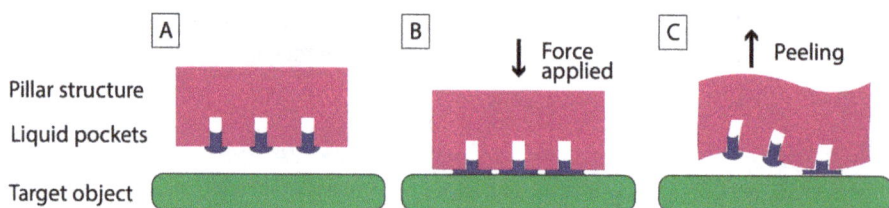

Figure 7.8: Example of capillary adhesion, where pockets of liquid are trapped in between pillar structures. When pressed against a target object, the pockets of liquid deform and create a liquid film between the pillars and the target, causing adhesion to occur. The pillars can buckle to break the film, reducing the adhesive force between the pad and target.

7.6 Example laboratory session A: surface textures

In this laboratory session, students will make textured surfaces and evaluate the impact of surface texture on the adhesion ability. The goal of the lab is to give them an understanding of the impact of surface texture on the adhesion ability across identical elastomers.

The materials required to make and test textured surfaces for improved adhesion are listed in Figure 7.9 and include:
- elastomer part A,
- elastomer part B,
- sandpaper of various grit (pictured are 150, 100, 60, and 40),
- mixing cup with a stir stick,
- filter paper,

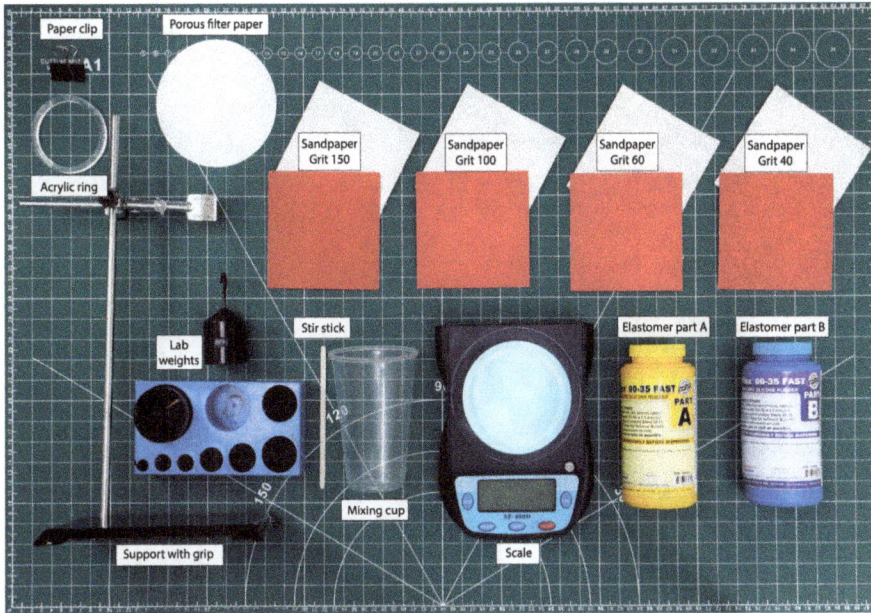

Figure 7.9: Required components for the assembly of textured surfaces with varying degrees of adhesion.

- acrylic ring,
- scale,
- paper clips,
- laboratory weights, and
- laboratory stand with grip.

The steps to follow in the production of the textured surface elastomers are shown in Figures 7.10 and 7.11. The steps to determine adhesion strength are listed separately in Figure 7.12.

1. Measure elastomer A.
2. Measure elastomer B.
3. Mix A and B.
4. Prepare mold by placing an acrylic ring on top of a sandpaper sheet with the textured surface facing up.
5. Pour elastomer mixture into mold.
6. Place filter paper in contact with the top uncured elastomer surface.
7. Cure at room temperature and remove from mold.
8. Secure the edge of the filter paper with the lab support grip.

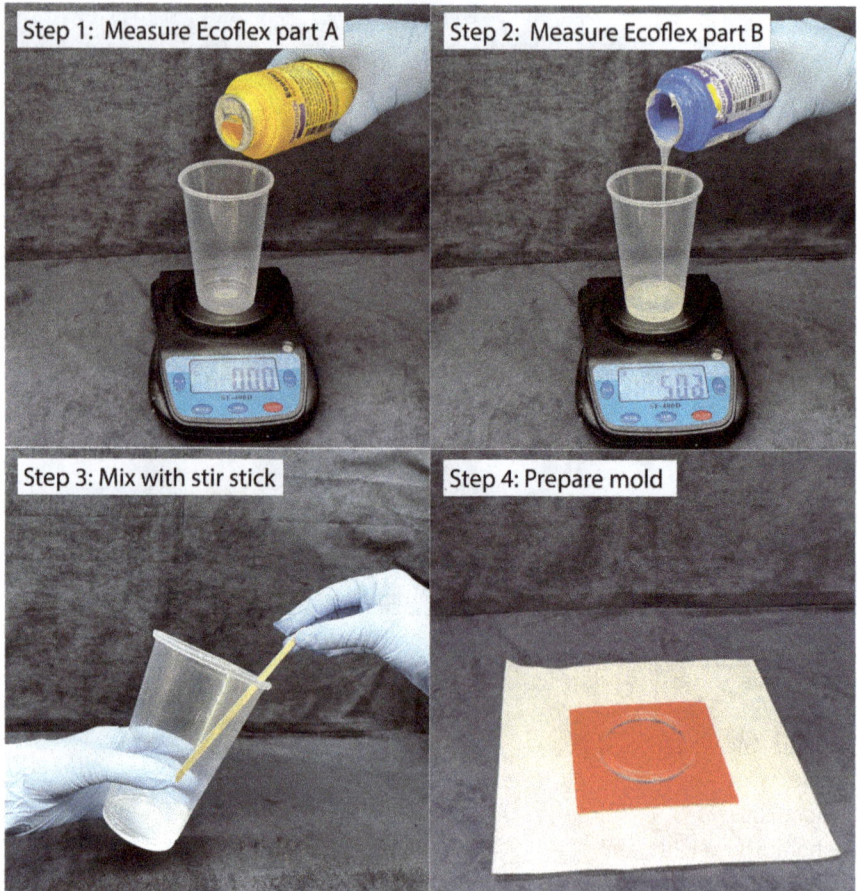

Figure 7.10: Steps 1–4 in the production of textured surfaces for adhesion.

9. Press a glass slide onto the surface and attach a paper clip at the bottom of the glass slide. Care should be taken to apply the same level of force on the glass slide for each adhesive pad tested.
10. Add weights and record how much each type of textured surface can hold.

For the project report, the students should plot the adhesion force as a function of surface grit. More than one elastomer should be tested to help the students evaluate the impact of elastomer properties compared to the surface texture of each material sample.

Step 5: Pour mix into mold

Step 6: Set paper on uncured mix

Step 7: Demold cured elastomer

Step 8: Secure on lab stand

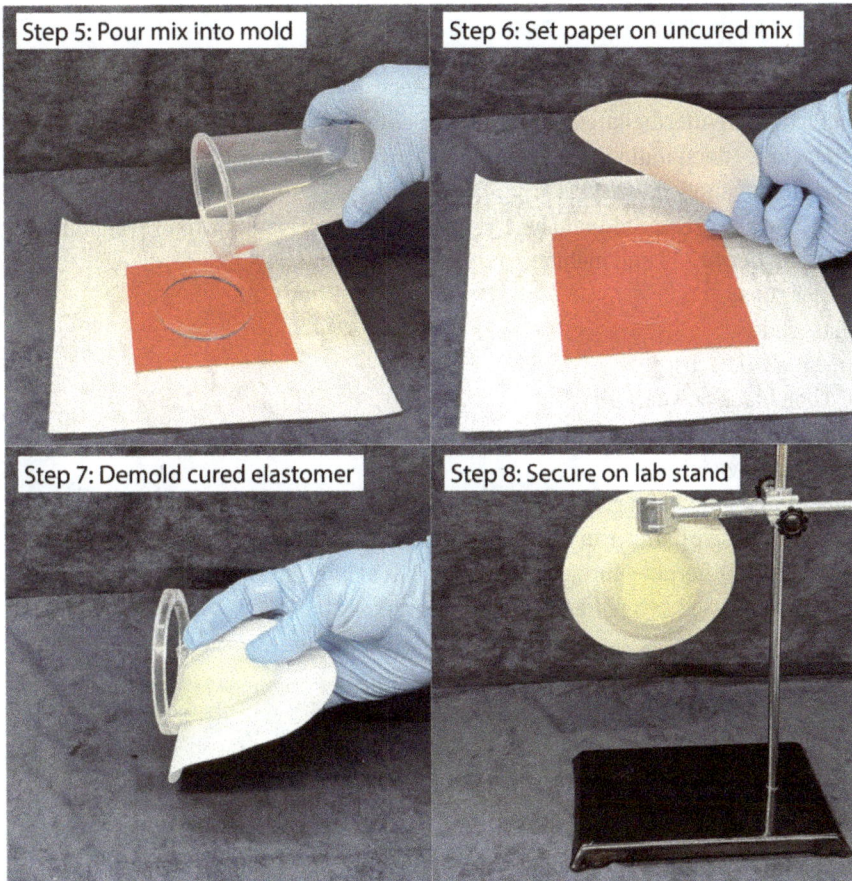

Figure 7.11: Steps 5–8 in the production of textured surfaces for adhesion.

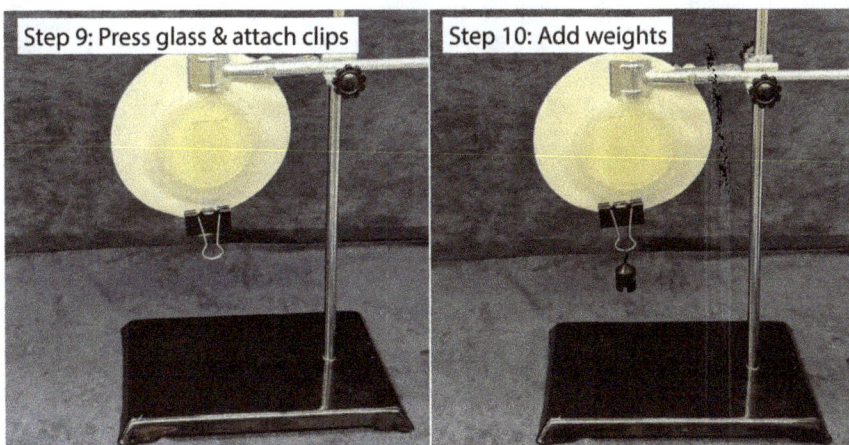

Step 9: Press glass & attach clips

Step 10: Add weights

Figure 7.12: Testing of adhesion ability via attachment of weights.

7.7 Example laboratory session B: electro-adhesives

In this laboratory session, students will make electro-adhesives and evaluate the impact of conductor and interdigitated pattern on the adhesion ability. The goal of the lab is to give them an understanding of the impact of design parameters on the adhesion ability as a function of applied voltage.

The materials required to make and test textured surfaces for improved adhesion are listed in Figure 7.13 and include:
- paper clips,
- glass slide,
- mylar as dielectric,
- carbon black powder,
- swab for spreading carbon black,
- laboratory weights,
- laboratory stand with grip, and
- electroadhesive pattern: in this example, laser cut siliconized mylar attached to VHB film. **Note:** Different subtractive methods have different resolutions; blade cutters may also be used as an alternative to laser cutters.

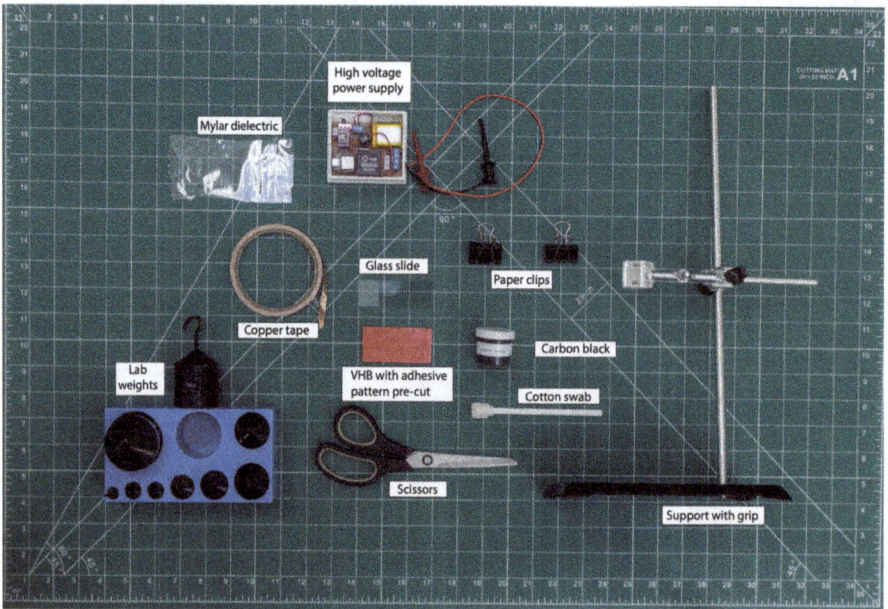

Figure 7.13: Required components for the assembly of an electro-adhesive and testing it in contact with a glass slide.

The steps to follow in the production of the electro-adhesive are shown in Figures 7.14 and 7.15. The steps to determine adhesion strength are listed separately in Figure 7.16.

1. Peel off the interdigitated electrodes from the VHB film (**Step 1**).
2. Spread carbon black directly onto VHB (**Step 2**).
3. Peel off the remaining mask (**Step 3**).
4. Attach copper tape to one side of the interdigitated pattern (**Step 4**).
5. Attach second tape (**Step 5**).
6. Press mylar dielectric onto the entire surface, pressing to ensure good contact and minimal air gaps (**Step 6**).
7. Trim excess dielectric (**Step 7**).

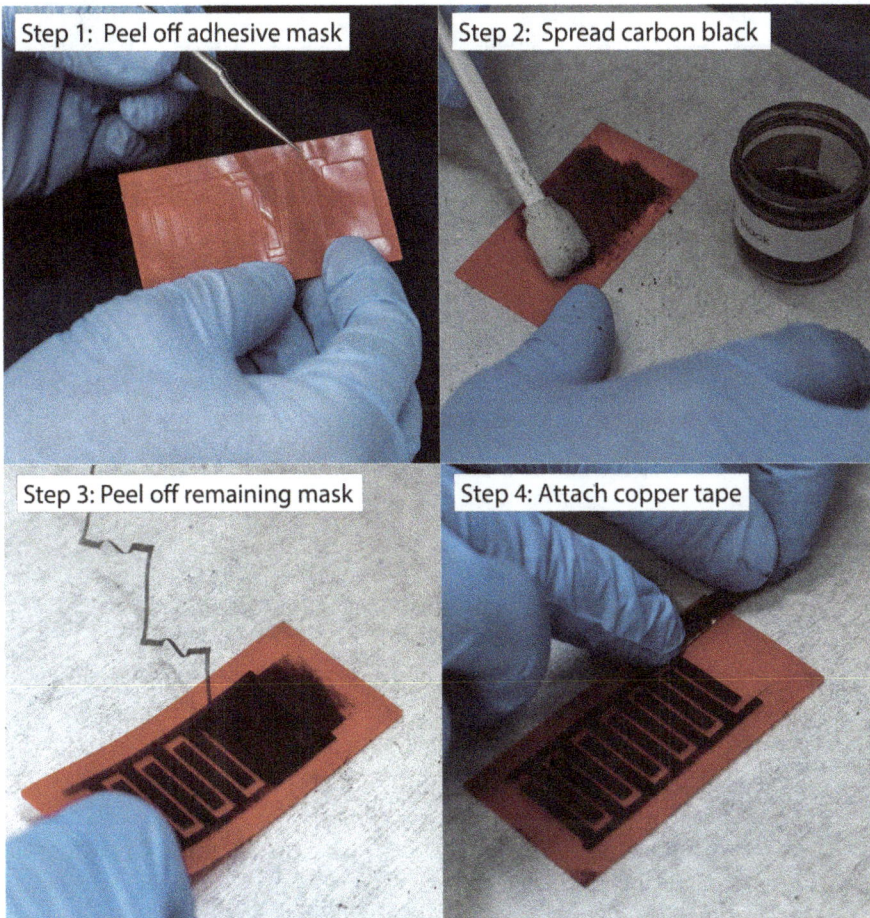

Figure 7.14: Steps 1–4 in the production of electro-adhesives.

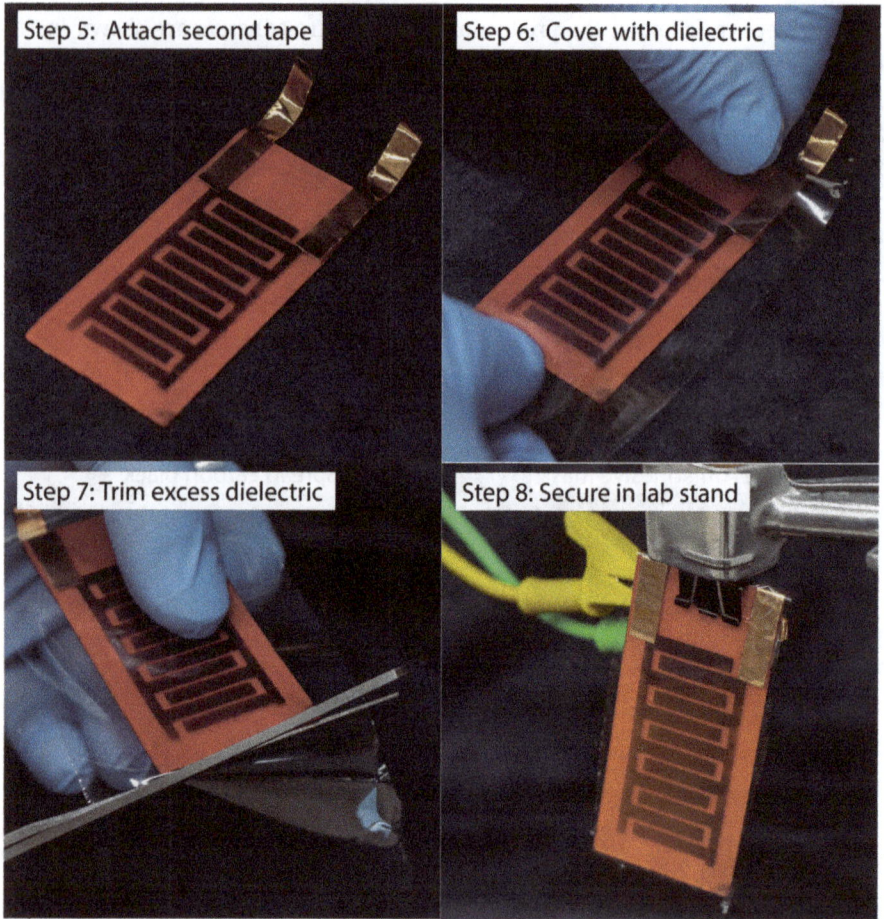

Figure 7.15: Steps 5–8 in the production of electro-adhesives.

8. Secure the edge of the adhesive with the lab support grip (**Step 8**). Connect to a high-voltage power supply and power on.
9. Press a glass slide onto the surface when the electro-adhseive is activated and attach a paper clip at the bottom of the glass slide. Care should be taken to apply the same level of force on the glass slide for each adhesive pad being tested (**Step 9**).
10. Add a paper clip and weights, then record each applied voltage (**Step 10**).

For the project report, the students should plot the adhesion force as a function of applied voltage for comparable adhesive pad sizes and patterns. More than one dielectric should be tested to help the students evaluate the impact of dielectric properties compared to the applied voltage.

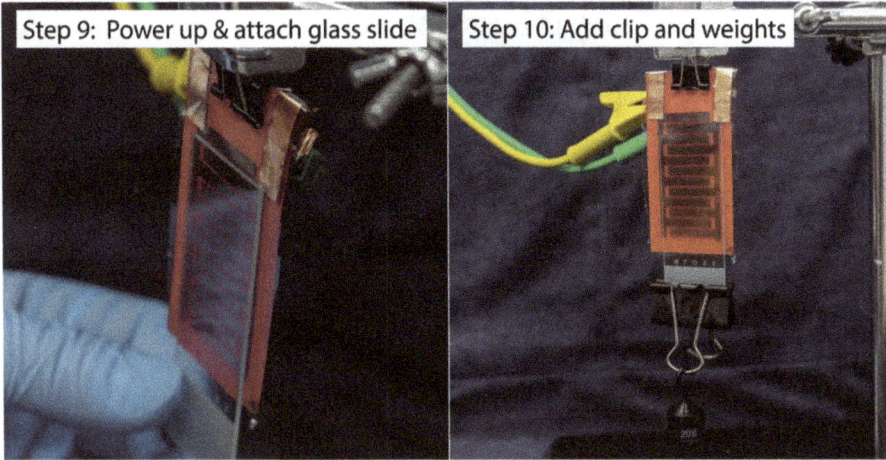

Figure 7.16: Testing of electro-adhesion via attachment of weights.

7.8 Example problems

Problem 1. A purely electrostatic adhesive consists of a set of flat electrodes covered by a Kapton dielectric layer of thickness 25 μm (25×10^{-6} m) and dielectric constant $\kappa =$ 3.5. The adhesive has a contact area of 10 cm^2 and is operated at an applied voltage of 3 kV. Using the parallel-plate capacitor model for electro-adhesion, and equation (7.1), determine the generated electrostatic pressure between the adhesive and surface and the total electro-adhesive force exerted by the pad.

Solution. We are given the voltage applied $V = 3000$ V, the Kapton dielectric constant $\kappa = 3.5$, the vacuum permittivity $\varepsilon_0 = 8.85 \times 10^{-12}$ F/m, the Kapton thickness $d = 25 \times 10^{-6}$ m, the adhesive area $A = 10$ cm$^2 = 10 \times 10^{-4}$ m^2.

Next we have to find the electrostatic pressure. We are given that the electrostatic adhesion follows the parallel-plate capacitor model, where the electrostatic pressure P is given by

$$P = \frac{1}{2}\varepsilon_r \varepsilon_0 \left(\frac{V}{d}\right)^2.$$

Substituting the values, we have

$$P = \frac{1}{2}(3.5)(8.85 \times 10^{-12})\left(\frac{3000}{25 \times 10^{-6}}\right)^2$$

$$= \frac{1}{2}(3.5)(8.85 \times 10^{-12})(1.2 \times 10^8)^2$$

$$= \frac{1}{2}(3.5)(8.85 \times 10^{-12})(1.44 \times 10^{16})$$

$$= \frac{1}{2}(3.5 \times 1.276 \times 10^5)$$

$$= \frac{1}{2}(4.466 \times 10^5)$$

$$= 2.233 \times 10^5 \text{ N/m}^2.$$

So the electrostatic pressure is 223.3 kPa.

From this we can find the total electro-adhesive force:

$$F = P \times A$$

$$= (2.233 \times 10^5) \times (10 \times 10^{-4})$$

$$= 22.33 \text{ N}.$$

Problem 2. For an adhesive pad shown in Figure 7.3, assume that there are 100 digits, with a spacing of 100 µm, a digit length of 1 cm, and a digit width of 400 µm. Estimate the total effective area of the electro-adhesive, i. e., the area of the digits that are useful for adhesion, ignoring the impact of connecting segments on the sides. Answer the following questions:

1. What is the total area of the adhesive pad if the electrode connectors are offset by 1 mm on either side, each 1-mm wide?
2. Using equation (7.1), what is the electro-adhesive force when a 5-kV signal is applied to an electro-adhesive with a 50-µm dielectric and a dielectric constant of 5?
3. Develop a more precise fabrication method, making digits of 100-µm width. If the total stays the same, how many more digits can you include? Does the equation predict a force increase or decrease? Does your intuition agree with the prediction?

Bibliography

Guoying Gu et al. "Soft wall-climbing robots". In: *Science Robotics* 3.25 (2018), eaat2874.

Wei Wang, Yang Liu, and Zongwu Xie. "Gecko-like dry adhesive surfaces and their applications: a review". In: *Journal of Bionic Engineering* (2021), pp. 1–34.

Jinsheng Zhao, Neng Xia, and Li Zhang. "A review of bioinspired dry adhesives: from achieving strong adhesion to realizing switchable adhesion". In: *Bioinspiration & Biomimetics* (2024), IOP Publishing.

8 Advanced topics in soft robotics

Contents

The topics covered in this chapter focus on a range of robotic functionalities, from additional actuation mechanisms to energy storage and logic methods required for control. In general, the topics are considered advanced because prototyping these types of machines requires more specialized equipment and infrastructure than the previous examples from the first seven chapters. This chapter does not include guidance for practical sessions, but still aims to present the key concepts from an energy perspective.

8.1 Deformable supercapacitors: ionic polymer–metal composites

In contrast to dielectric elastomer actuators (DEAs), which are parallel plate pure electrostatic capacitors, ionic polymer–metal composites (IPMCs) are supercapacitors that rely on ion movement to produce deformation. IPMCs are considered electroactive polymers (EAPs) that function as both soft actuators and sensors by exploiting the movement of ions and water molecules in response to an applied electric field.

The operation mode of IPMCs is similar to that of DEAs, but the operating voltage range is significantly lower (1–5 V). IPMCs consist of a polymer membrane (e. g., Nafion, Flemion, or Aquivion) sandwiched between two thin layers of metal electrodes (e. g., platinum, gold, or silver). The sandwich is flooded with an electrolyte, either an aqueous solution or an ionic liquid as shown in Figure 8.1. When voltage (≈1–5 V) is applied across the electrodes, cations inside the membrane migrate toward the cathode, carrying water molecules with them. This ion migration causes swelling on one side and contraction on the other to accommodate the ion size mismatch, leading to bending motion of the IPMC strip. When the voltage is reversed, ions start flowing in the opposite direction, and the bending direction switches.

The fabrication method for IPMCs requires assembly from the central membrane, which is critical for device operation. The sequence of steps includes:

– *Polymer membrane preparation:* The base material is usually Nafion, a perfluorosulfonic acid polymer that allows ion transport. The polymer sheet is pretreated in chemical baths (e. g., acid and base solutions) to remove impurities and improve ion exchange properties.

https://doi.org/10.1515/9783111069418-008

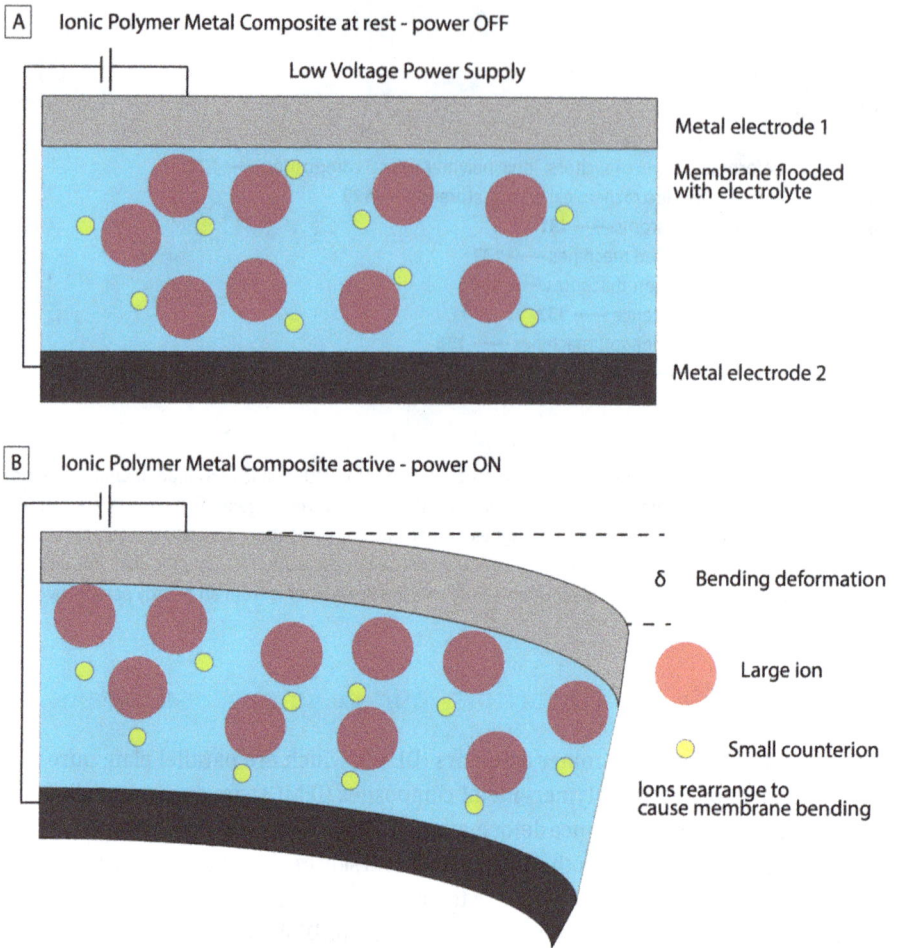

A Ionic Polymer Metal Composite at rest - power OFF

Low Voltage Power Supply

Metal electrode 1

Membrane flooded
with electrolyte

Metal electrode 2

B Ionic Polymer Metal Composite active - power ON

δ Bending deformation

Large ion

Small counterion

Ions rearrange to
cause membrane bending

Figure 8.1: Schematic of the operation of an ionic polymer metal composite made from sandwiching a membrane flooded with electrolyte between two metal electrodes. When a low voltage is applied, charges migrate in the supercapacitor. If the ions in electrolyte are mismatched in size, then the membrane bends to accommodate the ions rearrangement.

- *Electrode deposition:* Thin metal electrodes are deposited onto both surfaces of the polymer membrane via chemical and electroless plating or physical vapor deposition.
- *Electrolyte injection:* The IPMC is soaked in an electrolyte (e. g., lithium chloride, sodium chloride, or potassium chloride solutions in water) or an ionic liquid to ensure proper ion conductivity.
- *Integration into soft machines:* the IPMC is cut into desired shapes (strips, discs, or more complex geometries), and electrical connections are established using silver epoxy or conductive adhesives.

Contrasting with other soft actuators, IPMCs have some significant advantages, including operation at low voltage, which is easier to produce than the high-voltage requirement of dielectric elastomer actuators. However, widespread adoption of this method is poor because of several drawbacks, including electrolyte dependence, as the performance decreases in dry environments if the electrolyte dries out. Additionally, IPMCs generate low forces and can undergo electrode delamination under repeated use. Lastly, their operation as supercapacitors causes slow response times as charges need to migrate for motion to occur. Overall, the systems are still of significant interest in research settings but have limited adoption outside of the laboratory.

8.2 Deformable electrochemical energy storage

Many of the examples of autonomous robots in this book require on board sources of electricity, and almost all of them use commercially available rigid rechargeable batteries. Replacing all of the conventional battery components with deformable counterparts is a challenging endeavor because of the complex balance of material properties required for both stretchability and electrochemical stability. Several solutions have been demonstrated, and typically trade-off between battery performance, for example, by reducing energy density to achieve the ability to stretch under an applied strain.

For a deeper understanding, it is valuable to consider the components of a conventional lithium ion battery and evaluate how each part could be converted into a stretchable component:

– *Active materials:* the cathode materials are typically ceramic compounds (e.g., $LiFePO_4$, $LiMn_2O_4$, etc.), whereas the anode is most often graphite. These powders are mixed with carbon particles to increase electronic conductivity and binders such as polyvinilydene difluoride (PVDF) to retain electrode integrity, and flooded with electrolytes to provide ionic conductivity.
– *Electrolyte:* ion-rich solutions with high conductivity are required to allow for fast charging and discharging. High-voltage batteries operating above 2 V, operate outside the electrochemical stability window of water and require nonaqueous solutions, such as $LiPF_6$ in propylene carbonate.
– *Separator:* membranes permeable to electrolyte that separate the anode and cathode and allow only ions to pass, not electrons. The membranes are porous structures, made of polyethylene, polypropylene, or other inert polymers.
– *Current collectors:* both the cathode and anode are stacked against current collectors, which allow electrons to flow in parallel with the ions, and do electrical work. For high-voltage batteries, the material of each current collector must be carefully selected to ensure electrochemical stability. In Li-ion batteries that use graphite as the anode, the preferred metal on the anode is copper, whereas that of the cathode is aluminum.

– *Battery case:* to ensure long-term stability of the electrochemical cell, oxygen and water need to be kept out of the battery. This is done by encasing the entire system in metalized pouches or metal cases, with just the current collector terminals connected to the outside.

One example, shown in Figure 8.2, establishes a design strategy for making a complete Li-ion battery in which all the components are fully stretchable up to 100 % beyond their original length (Chen et al., "Fully integrated design of a stretchable solid-state lithium-ion full battery"). The current collectors, which are metallic and rigid in conventional batteries, are replaced with poly(styrene)-block-poly(ethyleneran-butylene)-block-poly(styrene) (SEBS) blends with carbon black, carbon nanotubes, and coated with silver flakes for improved electrical conductivity. The separator membrane is replaced with a water-in-salt electrolyte in a deformable hydrogel. The entire system uses a water-based electrolyte, which allows $LiMn_2O_4$ to serve as the cathode and V_2O_5 to serve as the anode in an electrochemical couple with a voltage of between 0.7 and 1.5 V depending on state of charge. Even stretching the battery to 50 %, the reversible capacity is 28 mAh g^{-1} and an average energy density of 20 Wh kg^{-1} after 50 cycles at 120 mA g^{-1}. The level of materials science expertise required to produce this type of device is significant, which so far has limited widespread adoption of the technology.

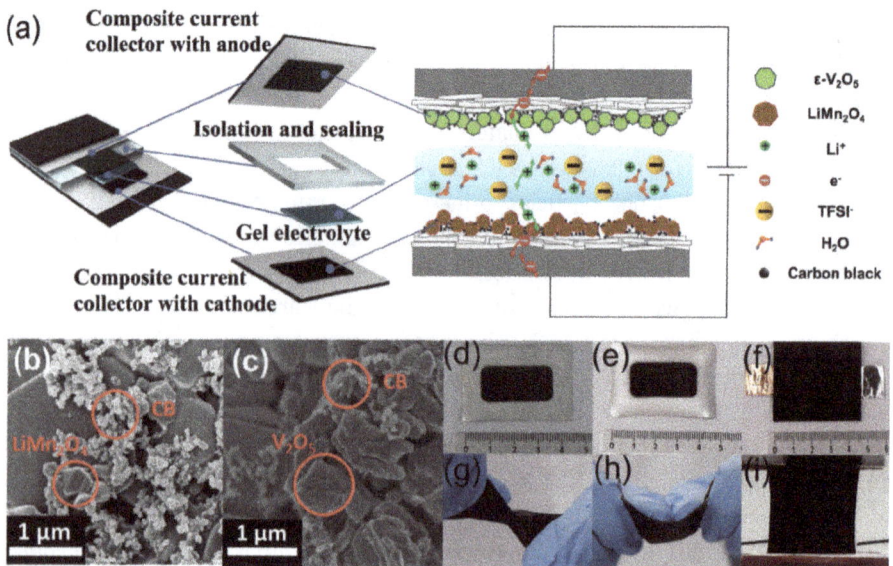

Figure 8.2: From Chen et al. 2019: Schematic illustration of the design and working principle of the stretchable full cell. b,c) SEM images of $LiMn_2O_4$ composite cathode V_2O_5 composite anode. d,e) Photos of the V_2O_5 composite anode and the $LiMn_2O_4$ composite cathode. f–i) Photos of the final full cell (f), under mechanical twisting (g), bending (h), and stretching (i).

8.3 Fluidic logic systems

Most often in soft robotics, the computing required for functionality is done by rigid microcontrollers. These components are not soft but have a small footprint relative to the rest of the robot body and can be integrated with relative ease. Alternatively, with fluid sources used for actuation available in a soft robot, it is possible to repurpose some of that infrastructure for enabling logic and control. Microfluidic logic circuits use fluid flow and pressure differences to perform logical operations, similarly to how electronic circuits use electrical signals. These circuits enable fully soft, self-contained, and pneumatic control systems for soft robots, eliminating the need for rigid electronics, batteries, or processors.

The operation of microfluidic logic systems relies on channels, valves, and pressure regulators to control fluid* in response to external inputs. Instead of electrical signals (0 s and 1 s), these systems encode logic using fluid pressure states:

– High pressure (1) → **Valve open** → Actuation occurs
– Low pressure (0) → **Valve closed** → No actuation

Just like electronic circuits, microfluidic systems can implement Boolean logic using soft valves and pneumatic channels:

– **AND Gate:** Two input pressures must be high for an output to be generated.
– **OR Gate:** Any one of the input pressures being high triggers an output.
– **NOT Gate:** A high-pressure input results in a low-pressure output (inverting the signal).
– **NOR Gate:** A negation of the OR gate, only allowing an output when a single input is active, not both.

These components can be combined to create complex sequential logic, enabling fully autonomous soft robot behaviors.

Figure 8.3 shows a research example of a pneumatic logic gate made via additive manufacturing, specifically 3D printing, which enables fluid control of soft machine (Conrad et al., "3D-printed digital pneumatic logic for the control of soft robotic actuators"). An example of a desired outcome is autonomously sequenced movements: without electronics, soft robots can be programmed to move in predefined sequences using fluidic timers and logic gates. For example, a soft walking robot can be driven by alternating pressurized fluid pulses, moving its legs without the need for microcontrollers, or other electronics susceptible to electromagnetic interference.

Some of the limitations of this approach have to do with fabrication and operation of the microfluidic logic systems. Fabricating microfluidic logic systems is a fairly complex process, which often requires precision microfluidic engineering in clean room environments. Once built, the fluidic logic systems have limited processing speed, an order of magnitude slower than electronic circuits because matter needs to move for the logic operations to be completed. Lastly, the entire system still relies on available pressure

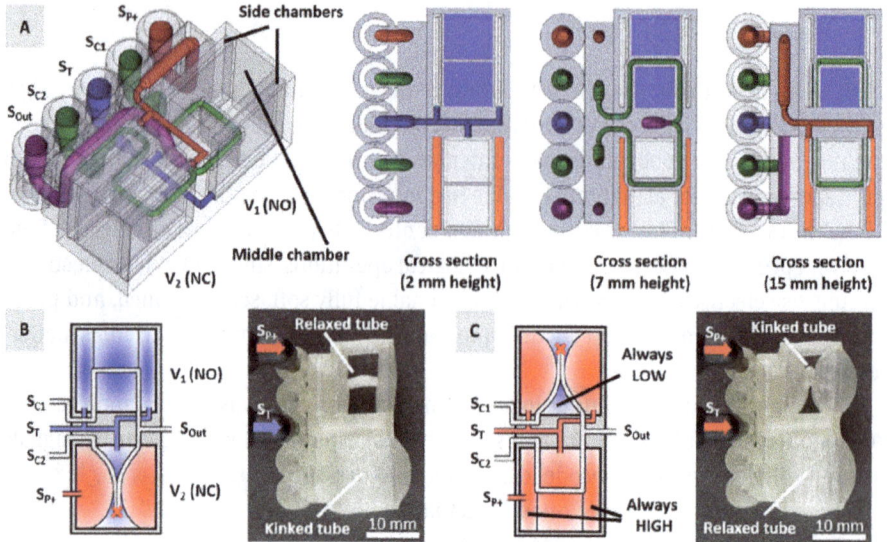

Figure 8.3: From reference Conrad et al., 2024, Pneumatic logic gate (PLG). (A) Model and cross-sections at varying heights of the PLG consisting of an NO valve (V_1) and an NC valve (V_2). Two sockets (S_{C1} and S_{C2}) supply input channels (green), leading through one valve each and merging in the output socket S_{Out} (purple). Because of a permanent pressure supply from S_{P+} (red), (V_2) works in NC mode, whereas V_1 is NO. Accordingly, they operate alternately, controlled by the toggle socket S_T (blue). (B, C) Schematic illustration and physical PLG 3D printed from TPU with different signals.

lines, which cause downstream integration challenges, at the interface with pumps or other fluid sources for continuous operation.

8.4 Osmosis-powered machines

Reverse osmosis (RO), a process typically used for water purification, can be repurposed to drive fluidic actuation in soft robots by exploiting osmotic pressure differences to generate movement. This approach is primarily plant inspired, as plants use ion transport through membranes to convert parts of their structures from flaccid to turgid. The most studied example is tendrils, which change curvature as they grow and curl around surrounding objects to give the plant additional support.

Figure 8.4 shows a schematic of the two end states of a robotic replica of a cell membrane capable of tuning the stiffness of a electrolyte-filled chamber via applied voltage. The chamber is flooded with an aqueous electrolyte and contains a membrane with carbon electrodes on a side of the membrane. When a low voltage is applied (<2 V), ions migrate to one side of the membrane, causing water to flow to the opposite side and changing the stiffness of the chamber. The bottom side of the image shows the impact of this change in *tendril-inspired robots*. Tendrils are slender thread-like appendages of

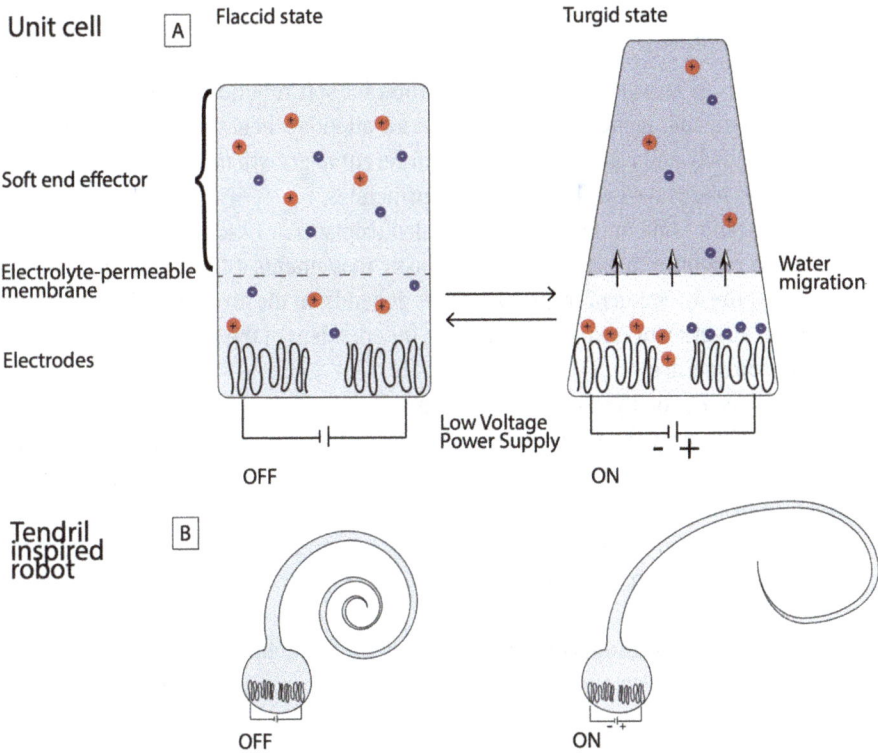

Figure 8.4: Top: Schematic of the deformation in an osmosis-powered actuator. The system contains an electrolyte-flooded chamber with an electrolyte permeable membrane and a set of electrodes on a side of the membrane. When the electrodes are applying a low voltage (<2 V), ions migrate to one side of a membrane, turning the opposite side turgid, allowing for soft robotic functionality. Bottom: deformation in a tendril inspired robot powered by osmosis, expanding an contracting a spiral tendril.

climbing plants, often growing in a spiral form, stretching out and twining around any suitable support. The change in stiffness also pushes water into the chamber and causes the robotic tendril to uncurl, reaching closer to nearby supports. When the voltage is removed, the ions are redistributed in the chamber, and the tendril returns to its curled state, anchoring better relative to its environment.

Some of the challenges of this approach are related to response speed, as osmotic flow is gradual and slow compared to traditional actuators. In addition, these systems produce limited force, so the anchoring force emerges more from the structure by curling a large surface area of the tendril and relying on local friction to support the rest of the robot.

8.5 Plant-inspired soft machines

Plants have served as a rich source of inspiration for soft robotics, particularly in their growth mechanisms, movement strategies, and adaptive behaviors. Unlike animals, plants move slowly and rely on hydraulic or differential growth mechanisms to achieve motion, which aligns well with soft robotics principles.

Plants exhibit some unique behaviors called tropisms or nastic movements, both of which are in response to external stimuli. Nastic movements differ from tropic movements in that the direction of tropic responses depends on the direction of the stimulus, whereas the direction of nastic movements is independent of the stimulus' position. Examples of tropisms are:

- *Phototropism*, growth or movement towards light.
- *Gravitropism*, growth or movement influenced by gravity, either toward it or opposite of it.
- *Skototropism*, growth or movement away from light.
- *Thigmotropism*, growth in response to touch, as seen in climbing plants, toward the location of the touch stimulus.

If the direction of the stimulus does not influence the behavior, then the responses are:

- *Thigmonasty or seismonasty* is the nastic (nondirectional) response of a plant or fungus to touch or vibration.
- *Nyctinasty*, circadian rhythm-based nastic movement of higher plants in response to the onset of darkness.
- Other nastic responses are possible to other stimuli, such as temperature (thermonasty), chemical species (chemonasty), etc.

From a soft robotics perspective, the most relevant mechanisms are responses to environmental changes, such as variations in relative humidity, and responses to external stimuli. In plants, *humidity-responsive materials* are often dead tissues, and the mechanism for reaction is cell wall swelling and shrinking. The behavior is seen is the desert resurrection plant, erodium awns, wheat seeds, dandelion pappi, pine cones, ice plant seeds capsules, and *Bauhinia variegate* seedpods. In contrast, mechanically responsive materials are living cells, and the response is due to changes in *turgor pressure*.

An example of *hydronasty* is the behavior of the dandelion (*Taraxacum officinale*), specifically the pappus, the umbrella-like structure attached to seeds (Meng et al., "Hydroactuated configuration alteration of fibrous dandelion pappi: Toward self-controllable transport behavior"). This structure exhibits a humidity-responsive behavior, where its filaments open in dry conditions and close in humid environments. This response is driven by hygroscopic movement, where changes in moisture cause differential expansion and contraction in plant tissues. Figure 8.5 shows an example from literature, showing that the pappus consists of a circular array of filaments connected at a central hub. In dry air, the filaments spread open, optimizing wind dispersal, whereas

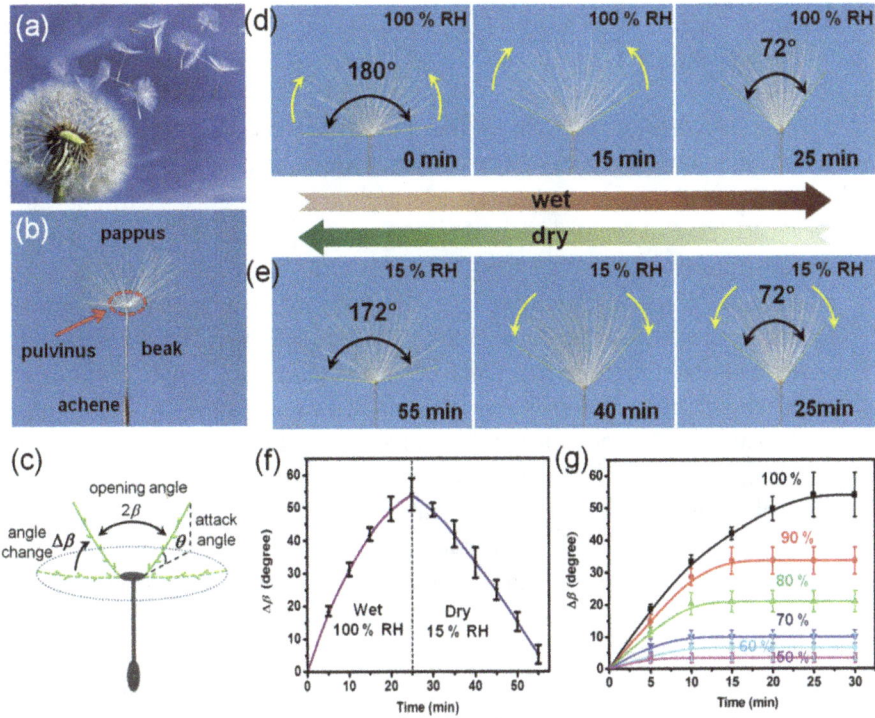

Figure 8.5: From Meng et al., 2016. Top: deformation of the pappus structure of a dandelion as the relative humidity increases. Bottom: changes in angle of attack over time and humidity range to control drag of the entire structure to allow for dissipation via wind to distribute seeds.

in high humidity, the filaments close to reduce air resistance and settle the seed. This motion is due to asymmetric swelling and contraction of the pappus tissue, where one side absorbs moisture faster than the other, causing bending.

An example of *thigmonasty* is the *Mimosa pudica*, which when touched rapidly, folds its leaves and droops its stems, likely to deter herbivores by appearing smaller and exposing its sharp spines. This movement is driven by changes in turgor pressure, the water pressure inside plant cells. High turgor pressure keeps cells rigid, whereas water loss makes them flaccid due to osmosis, the movement of water across membranes based on ion concentration gradients. In the pulvinus, a hinge-like structure at the leaf base, tactile stimulation triggers shifts in potassium and chloride ion concentrations between flexor and extensor cells. Water moves from the extensor (top) cells to the flexor (bottom) cells, causing the extensor cells to lose turgor and collapse, whereas the flexor cells swell, folding the leaflets and drooping the midrib. The folding occurs within 4–5 seconds, whereas unfolding takes up to 10 minutes.

Plants achieve fast motion through three distinct biomechanical strategies: bistable snap-through mechanisms, fracture release mechanisms, and cavitation release mechanisms. Each mode relies on different physical principles to store and release energy efficiently, allowing for rapid movement despite plants' lack of fast moving actuators.

1. *Bistable snap-through mechanisms:* this mode relies on elastic instabilities where stored mechanical energy is released suddenly when a structure shifts between two stable states. Plants accumulate elastic energy in a prestressed structure, which, upon triggering, undergoes a snap-through transition, producing rapid movement. The classic example is the Venus flytrap (*Dionaea muscipula*), where the trap's lobes are naturally curved in a convex shape when open. Stimulation of trigger hairs alters the internal turgor pressure, shifting the lobes past a critical threshold. This causes an instantaneous snap-through inversion, closing the trap in about 100 milliseconds, efficiently capturing prey with minimal energy loss.

2. *Cavitation release mechanisms:* cavitation-based motion occurs when negative pressure in xylem vessels or specialized water-filled cells leads to the sudden formation of vapor bubbles (cavitation), which results in a rapid collapse and energy release. A known example is the spore dispersal mechanisms of ferns (*Polypodiaceae*): their spore cases rely on a row of water-filled cells in the annulus. As these cells lose water due to evaporation, tension increases until cavitation occurs. This instantaneous bubble formation collapses the annulus, snapping it forward and catapulting spores into the air. Figure 8.6 captures a literature example of the mechanism.

3. *Fracture release mechanisms:* plants can generate rapid motion by using mechanical fracture to release stored elastic energy. In this strategy, a plant structure holds energy in a stressed or locked position, which is suddenly released when a weak point or fracture zone gives way. An example of explosive seed dispersal in *Impatiens* (or Touch-me-not plant) relies on tension that builds in the fruit pod walls as

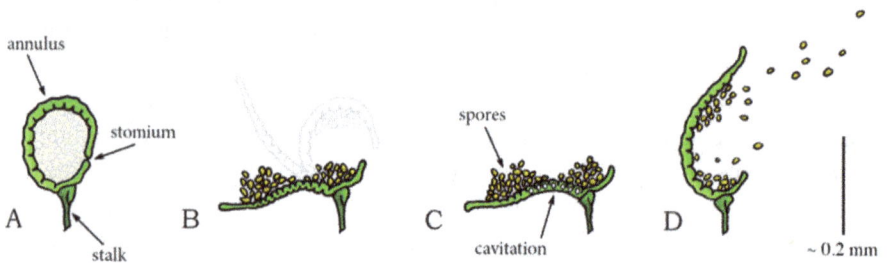

Figure 8.6: From Sakes et al., 2016: Cavitation catapult mechanism in the family Polypodiaceae or common ferns. (A) The mature sporangium in common ferns consisting of a stalk and an annulus enclosing the spores. (B) Dehydration of the annulus cells causes the radial cell walls to come closer together and the lateral walls to collapse internally, straightening the annulus. (C) When a critical pressure (between −9 and −20 MPa relative to ambient) is reached, cavitation occurs in the cells of the annulus. (D) Discharge of the spores by quick release of the elastic energy stored in the cell walls as the annulus snaps back to its original shape.

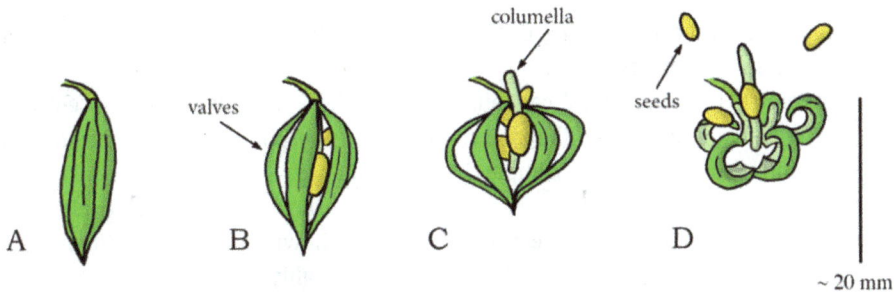

Figure 8.7: From Sakes et al., 2016: Swelling coiling catapult mechanism in Impatiens capensis. (A) The seedpod consisting of five interconnected valves. Elastic energy is stored in the seedpod by the absorption of water in the valves. When a critical pressure is reached, dehiscence of the valves from the columella and subsequent coiling discharges the seeds (A–D). (A) Shows the situation at $t = 0$ ms. Duration from (A) to (D) lasts about 3 to 4 ms.

they dry out (Sakes et al., "Shooting mechanisms in nature: a systematic review"), Figure 8.7 captures a literature example of the mechanism. When the structural integrity weakens beyond a threshold, the pod fractures and curls back rapidly, ejecting seeds at high speed.

As in other bioinspired examples, a deep understanding of the biological mechanisms allows researchers to build robotic replicas that capture the fundamental operation mechanism. Both passive and active movements inspired by plants have been reproduced in a range of robotic demonstrations. Lastly, plants grow by cell elongation and division, enabling them to adapt to their environment. Soft robots inspired by this principle use additive material deposition or inflation-based extension to move. One example has already been presented at the end of Chapter 3: eversion machines that grow by unrolling a thin tube from the tip, mimicking root growth in search of nutrients. Another example was shown earlier in this chapter, where osmosis is controlled by electrical signals to modify turgor pressure in a robotic tendril.

8.6 Electroluminescence

Just as soft materials can be used for actuation, sensing, and computation, they can also be adapted for communication, in particular via bioinspired luminescence. Many natural systems use luminescence for communication, predation, distraction, or defense. Bioluminescent organisms, such as fireflies (*Lampyridae*) and deep-sea fish, emit light through chemical reactions to attract mates, signal warnings, or lure prey. In marine environments, anglerfish (*Melanocetus johnsonii*) use a luminescent lure to deceive and attract smaller fish, whereas squids, octopods, and cuttlefish show complex methods of flashing light to confused predators. This ability to produce and control light provides evolutionary advantages, helping organisms blend into their surroundings through counterillumination or deter threats by mimicking more dangerous species.

Inspired by these biological capabilities, researchers have demonstrated electroluminescence using materials like zinc sulfide (ZnS) phosphorus-doped dielectrics to achieve similar functions. One example (Larson et al., "Highly stretchable electroluminescent skin for optical signaling and tactile sensing") shows hydrogel electrodes surrounding a zinc sulfide-doped dielectric elastomer, as shown in Figure 8.8. When powered at 700 Hz and 2.5 V per micron of dielectric, these highly stretchable systems show light-emitting behavior. The system can be adapted into displaying multiple pixels and into producing light of different wavelength, depending on the type of dopant in the zinc sulfide material.

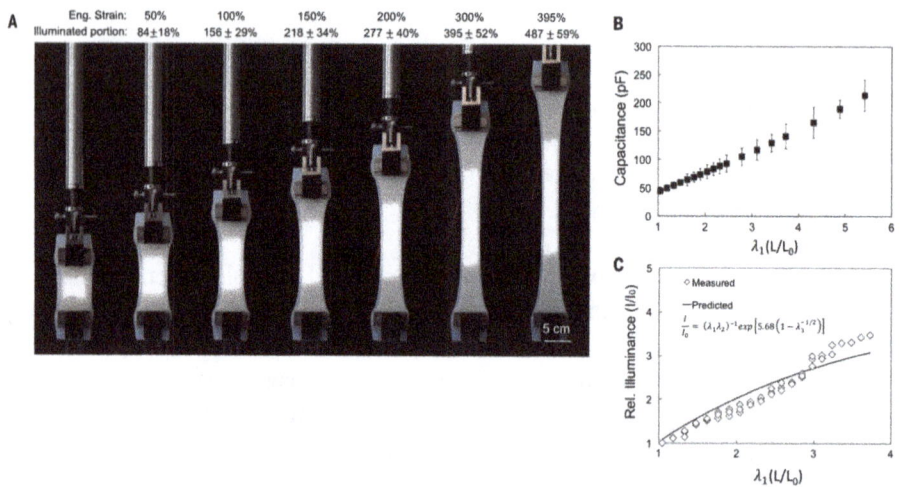

Figure 8.8: From Larson et al., 2016: An example elastomer which maintains its ability to produce light even when undergoing strain near 500 % compared to its original lenght. B. Capacitance and C. Relative illuminance as a function of the applied strain.

8.7 Chemical fuels for soft machines

Chemical fuels for soft machines have one distinct advantages over all other approaches: chemical fuels have high specific energy (in J/kg), higher than mechanical or electrochemical energy storage approaches. Additionally, natural muscles convert chemical fuels to mechanical work, and matching that ability can enable novel bioinspired robot designs. Unfortunately, to date, no chemical fuel-powered actuators can match the range of motion, specific energy, and response speed of natural muscles.

Some of the approaches that have been tried include:
1. *Combustion powered machines*, already described in Chapter 3. This approach ensures rapid combustion of chemical fuels such as methane or butane to produce

gas expansion for movement. An advantage of this method is that it provides high-power actuation without external pumps or motors.

2. *Fuel decomposing machines*, which rely on a simple mechanism hydrogen peroxide (H_2O_2) decomposition to release oxygen and heat. The oxygen gas produced expands soft chambers for actuation. This approach requires platinum or manganese dioxide catalysts to break down H_2O_2 inside elastomeric chambers and enables untethered operation, as no external compressors are needed. An alternative use of these fuels is to have the machine operate in the fuel itself. Asymmetric particles can preferentially produce gas on one side, which propels the particle forward within the fuel medium. A significant limitation of this approach is the need to operate in the fuel, which in the case of hydrogen peroxide is a strong oxidant and damaging to a range of materials.

3. *Localized reactions in hybrid actuators.* Localized oxidation of fuels can produce heat that causes other deformation to occur. In the example shown in Figure 8.9, methanol is oxidized on a platinum catalyst to produce heat, which in turn causes a shape memory alloy wire to contract (Yang, Chang, and Pérez-Arancibia, "An 88-milligram insect-scale autonomous crawling robot driven by a catalytic artificial muscle"). That contraction moves the legs of a autonomous microrobot, which requires no electronics for full energy autonomy.

Finally, it is worth summarizing the mechanism by which natural muscles operate, as they are the aspirational goal of soft roboticists in any novel actuator exploration. Muscles generate movement by converting chemical energy into mechanical work through a series of biochemical and biophysical processes. The primary chemical fuel for mus-

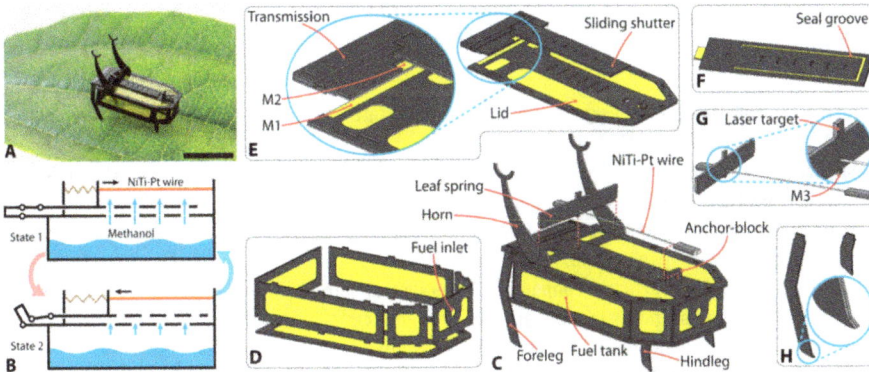

Figure 8.9: From Yang et al., 2020, (A) Photograph of a RoBeetle prototype resting on a leaf (the scale bar indicates a distance of 10 mm). (B) Schematic diagram of RoBeetle's actuation mechanism. (C) Exploded view of the robotic assembly. (D) Exploded view of the fuel tank subassembly. (E) Exploded view of the tank lid, transmission, and sliding shutter. (F) Bottom side of the sliding shutter. (G) NiTi-Pt composite wire and leaf spring. (H) Forelegs and hindlegs with bioinspired backward-oriented claws.

cle contraction is adenosine triphosphate (ATP), which powers the interaction between actin and myosin filaments in muscle fibers. ATP is a high-energy molecule that provides immediate energy for muscle contraction. When ATP is hydrolyzed by the enzyme myosin ATPase, it breaks down into adenosine diphosphate (ADP) and inorganic phosphate (Pi), releasing energy:

$$ATP \rightarrow ADP + Pi + Energy.$$

This energy is used to drive the cross-bridge cycle, allowing myosin heads to pull actin filaments, producing contraction. The cross-bridge cycle can be broken down into the following steps:

1. *ATP binding:* Myosin heads start in a low-energy state and bind to ATP, causing them to detach from actin.
2. *ATP hydrolysis:* ATP is broken down into ADP + Pi, causing the myosin head to shift into a high-energy "cocked" position.
3. *Cross-bridge formation:* The myosin head binds to actin, forming a cross-bridge.
4. *Power stroke:* Pi is released, and the myosin head pulls actin forward, shortening the muscle.
5. *ADP release:* The myosin head releases ADP, returning to its original state. A new ATP molecule binds, restarting the cycle.

The size of the myosin and actin filaments, which are in the 5–20 nm range allows these reactions to occur rapidly, powering fast movements as high as 300 Hz in flapping wing insects. Fabricating systems of the same size and functionality remains an open challenge, which will continue to drive research and innovation in Soft Robotics.

Bibliography

Xi Chen et al. "Fully integrated design of a stretchable solid-state lithium-ion full battery". In: *Advanced Materials* 31.43 (2019), p. 1904648.

Stefan Conrad et al. "3D-printed digital pneumatic logic for the control of soft robotic actuators". In: *Science robotics* 9.86 (2024), eadh4060.

Christina Larson et al. "Highly stretchable electroluminescent skin for optical signaling and tactile sensing". In: *Science* 351.6277 (2016), pp. 1071–1074.

Qing'an Meng et al. "Hydroactuated configuration alteration of fibrous dandelion pappi: Toward self-controllable transport behavior". In: *Advanced Functional Materials* 26.41 (2016), pp. 7378–7385.

Aimée Sakes et al. "Shooting mechanisms in nature: a systematic review". In: *Plos one* 11.7 (2016), e0158277.

Xiufeng Yang, Longlong Chang, and Néstor O Pérez-Arancibia. "An 88-milligram insect-scale autonomous crawling robot driven by a catalytic artificial muscle". In: *Science Robotics* 5.45 (2020), eaba0015.

Index

https://doi.org/10.1515/9783111069418-009

www.ingramcontent.com/pod-product-compliance
Lightning Source LLC
Chambersburg PA
CBHW081538220326
41598CB00036B/6475